Glencoe McGraw-Hill

Homework Practice Workbook

Geometry

McGraw Hill Glencoe

To the Student

This *Homework Practice Workbook* gives you additional problems for the concept exercises in each lesson. The exercises are designed to aid your study of mathematics by reinforcing important mathematical skills needed to succeed in the everyday world. The materials are organized by chapter and lesson, with one Practice worksheet for every lesson in *Glencoe Geometry*.

To the Teacher

These worksheets are the same ones found in the Chapter Resource Masters for *Glencoe Geometry*. The answers to these worksheets are available at the end of each Chapter Resource Masters booklet.

The *McGraw-Hill* Companies

 Glencoe

Send all inquiries to:
Glencoe/McGraw-Hill
8787 Orion Place
Columbus, OH 43240

ISBN: 978-0-07-890849-1
MHID: 0-07-890849-3

Homework Practice Workbook, Geometry

Printed in the United States of America

25 26 27 LHS 17

Contents

1-1 Skills Practice

Points, Lines, and Planes

Refer to the figure.

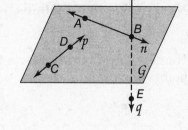

1. Name a line that contains point E.

2. Name a point contained in line n.

3. What is another name for line p?

4. Name the plane containing lines n and p.

Draw and label a figure for each relationship.

5. Point K lies on $\overleftrightarrow{RT}$.

6. Plane $\mathcal{J}$ contains line s.

7. $\overleftrightarrow{YP}$ lies in plane $\mathcal{B}$ and contains point C, but does not contain point H.

8. Lines q and f intersect at point Z in plane $\mathcal{U}$.

Refer to the figure.

9. How many planes are shown in the figure?

10. How many of the planes contain points F and E?

11. Name four points that are coplanar.

12. Are points A, B, and C coplanar? Explain.

1-1 Practice

Points, Lines, and Planes

Refer to the figure.

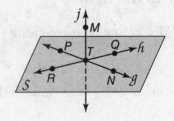

1. Name a line that contains points T and P.

2. Name a line that intersects the plane containing points Q, N, and P.

3. Name the plane that contains $\overleftrightarrow{TN}$ and $\overleftrightarrow{QR}$.

Draw and label a figure for each relationship.

4. $\overleftrightarrow{AK}$ and $\overleftrightarrow{CG}$ intersect at point M in plane $\mathcal{T}$.

5. A line contains $L(-4, -4)$ and $M(2, 3)$. Line q is in the same coordinate plane but does not intersect $\overleftrightarrow{LM}$. Line q contains point N.

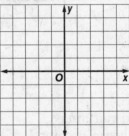

Refer to the figure.

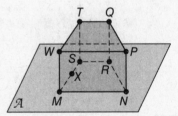

6. How many planes are shown in the figure?

7. Name three collinear points.

8. Are points N, R, S, and W coplanar? Explain.

VISUALIZATION Name the geometric term(s) modeled by each object.

9.

10. tip of pin→

11. strings

12. a car antenna

13. a library card

1-2 Skills Practice

Linear Measure

Find the length of each line segment or object.

1.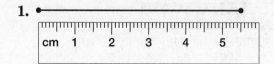

2.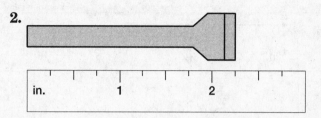

Find the measurement of each segment. Assume that each figure is not drawn to scale.

3. $\overline{NQ}$

Q ——1in.—— P ——$1\frac{1}{4}$ in.—— N

4. $\overline{AC}$

A ——4.9 cm—— B ——5.2 cm—— C

5. $\overline{GH}$

F ——9.7 mm—— G — H
|———— 15 mm ————|

ALGEBRA **Find the value of *x* and *YZ* if *Y* is between *X* and *Z*.**

6. $XY = 5x$, $YZ = x$, and $XY = 25$

7. $XY = 12$, $YZ = 2x$, and $XZ = 28$

8. $XY = 4x$, $YZ = 3x$, and $XZ = 42$

9. $XY = 2x + 1$, $YZ = 6x$, and $XZ = 81$

Determine whether each pair of segments is congruent.

10. $\overline{BE}$, $\overline{CD}$

B —2 m— C
3 m   3 m
E —— 5 m —— D

11. $\overline{MP}$, $\overline{NP}$

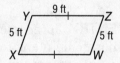

12. $\overline{WX}$, $\overline{WZ}$

Y —9 ft— Z
5 ft   5 ft
X   W

1-2 Practice

Linear Measure

Find the length of each line segment or object.

1.

2.

Find the measurement of each segment. Assume that each figure is not drawn to scale.

3. *PS*

P •——18.4 cm——• Q —4.7 cm— • S

4. *AD*

A •——$2\frac{3}{8}$ in.——• C —$1\frac{1}{4}$ in.— • D

5. *WX*

W • —X • ——89.6 cm—— • Y
|←———— 100 cm ————→|

ALGEBRA Find the value of *x* and *KL* if *K* is between *J* and *L*.

6. $JK = 6x$, $KL = 3x$, and $JL = 27$

7. $JK = 2x$, $KL = x + 2$, and $JL = 5x - 10$

Determine whether each pair of segments is congruent.

8. $\overline{TU}$, $\overline{SW}$

9. $\overline{AD}$, $\overline{BC}$

10. $\overline{GF}$, $\overline{FE}$

11. **CARPENTRY** Jorge used the figure at the right to make a pattern for a mosaic he plans to inlay on a tabletop. Name all of the congruent segments in the figure.

1-3 Skills Practice

Distance and Midpoints

Use the number line to find each measure.

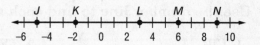

1. LN

2. JL

3. KN

4. MN

Find the distance between each pair of points.

5.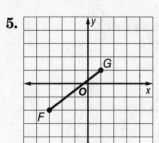

6.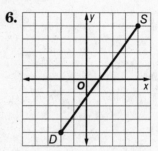

7. $K(2, 3)$, $F(4, 4)$

8. $C(-3, -1)$, $Q(-2, 3)$

9. $Y(2, 0)$, $P(2, 6)$

10. $W(-2, 2)$, $R(5, 2)$

11. $A(-7, -3)$, $B(5, 2)$

12. $C(-3, 1)$, $Q(2, 6)$

Use the number line to find the coordinate of the midpoint of each segment.

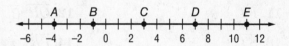

13. $\overline{DE}$

14. $\overline{BC}$

15. $\overline{BD}$

16. $\overline{AD}$

Find the coordinates of the midpoint of a segment with the given endpoints.

17. $T(3, 1)$, $U(5, 3)$

18. $J(-4, 2)$, $F(5, -2)$

Find the coordinates of the missing endpoint if P is the midpoint of $\overline{NQ}$.

19. $N(2, 0)$, $P(5, 2)$

20. $N(5, 4)$, $P(6, 3)$

21. $Q(3, 9)$, $P(-1, 5)$

1-3 Practice

Distance and Midpoints

Use the number line to find each measure.

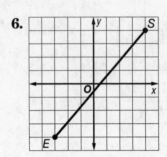

1. VW 2. TV

3. ST 4. SV

Find the distance between each pair of points.

5. 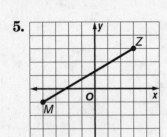   6.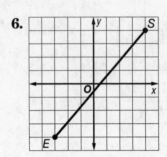

7. $L(-7, 0)$, $Y(5, 9)$ 8. $U(1, 3)$, $B(4, 6)$

9. $V(-2, 5)$, $M(0, -4)$ 10. $C(-2, -1)$, $K(8, 3)$

Use the number line to find the coordinate of the midpoint of each segment.

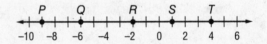

11. $\overline{RT}$ 12. $\overline{QR}$

13. $\overline{ST}$ 14. $\overline{PR}$

Find the coordinates of the midpoint of a segment with the given endpoints.

15. $K(-9, 3)$, $H(5, 7)$ 16. $W(-12, -7)$, $T(-8, -4)$

Find the coordinates of the missing endpoint if E is the midpoint of $\overline{DF}$.

17. $F(5, 8)$, $E(4, 3)$ 18. $F(2, 9)$, $E(-1, 6)$ 19. $D(-3, -8)$, $E(1, -2)$

20. **PERIMETER** The coordinates of the vertices of a quadrilateral are $R(-1, 3)$, $S(3, 3)$, $T(5, -1)$, and $U(-2, -1)$. Find the perimeter of the quadrilateral. Round to the nearest tenth.

1-4 Skills Practice

Angle Measure

For Exercises 1–12, use the figure at the right.

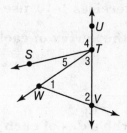

Name the vertex of each angle.

1. ∠4   2. ∠1

3. ∠2   4. ∠5

Name the sides of each angle.

5. ∠4   6. ∠5

7. ∠STV   8. ∠1

Write another name for each angle.

9. ∠3   10. ∠4

11. ∠WTS   12. ∠2

Classify each angle as *right*, *acute*, or *obtuse*. Then use a protractor to measure the angle to the nearest degree.

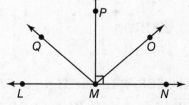

13. ∠NMP   14. ∠OMN

15. ∠QMN   16. ∠QMO

ALGEBRA In the figure, $\overrightarrow{BA}$ and $\overrightarrow{BC}$ are opposite rays, $\overrightarrow{BD}$ bisects ∠EBC.

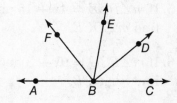

17. If $m\angle EBD = 4x + 16$ and $m\angle DBC = 6x + 4$, find $m\angle EBD$.

18. If $m\angle EBD = 4x - 8$ and $m\angle EBC = 5x + 20$, find the value of x and $m\angle EBC$.

1-4 Practice

Angle Measure

For Exercises 1–10, use the figure at the right.

Name the vertex of each angle.

1. ∠5   2. ∠3

3. ∠8   4. ∠NMP

Name the sides of each angle.

5. ∠6   6. ∠2

7. ∠MOP   8. ∠OMN

Write another name for each angle.

9. ∠QPR   10. ∠1

Classify each angle as *right*, *acute*, or *obtuse*. Then use a protractor to measure the angle to the nearest degree.

11. ∠UZW   12. ∠YZW

13. ∠TZW   14. ∠UZT

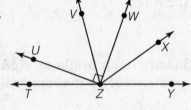

ALGEBRA In the figure, $\overrightarrow{CB}$ and $\overrightarrow{CD}$ are opposite rays, $\overrightarrow{CE}$ bisects ∠DCF, and $\overrightarrow{CG}$ bisects ∠FCB.

15. If $m\angle DCE = 4x + 15$ and $m\angle ECF = 6x - 5$, find $m\angle DCE$.

16. If $m\angle FCG = 9x + 3$ and $m\angle GCB = 13x - 9$, find $m\angle GCB$.

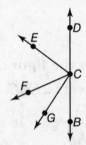

17. **TRAFFIC SIGNS** The diagram shows a sign used to warn drivers of a school zone or crossing. Measure and classify each numbered angle.

1-5 Skills Practice

Angle Relationships

For Exercises 1–6, use the figure at the right. Name an angle or angle pair that satisfies each condition.

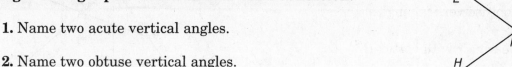

1. Name two acute vertical angles.

2. Name two obtuse vertical angles.

3. Name a linear pair.

4. Name two acute adjacent angles.

5. Name an angle complementary to ∠*EKH*.

6. Name an angle supplementary to ∠*FKG*.

7. Find the measures of an angle and its complement if one angle measures 24 degrees more than the other.

8. The measure of the supplement of an angle is 36 less than the measure of the angle. Find the measures of the angles.

ALGEBRA For Exercises 9–10, use the figure at the right.

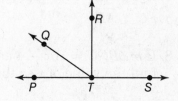

9. If $m\angle RTS = 8x + 18$, find the value of x so that $\overrightarrow{TR} \perp \overrightarrow{TS}$.

10. If $m\angle PTQ = 3y - 10$ and $m\angle QTR = y$, find the value of y so that ∠*PTR* is a right angle.

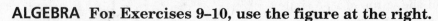

Determine whether each statement can be assumed from the figure. Explain.

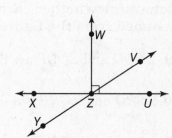

11. ∠*WZU* is a right angle.

12. ∠*YZU* and ∠*UZV* are supplementary.

13. ∠*VZU* is adjacent to ∠*YZX*.

1-5 Practice

Angle Relationships

Name an angle or angle pair that satisfies each condition.

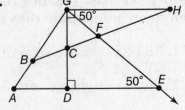

1. Name two obtuse vertical angles.

2. Name a linear pair with vertex B.

3. Name an angle not adjacent to, but complementary to $\angle FGC$.

4. Name an angle adjacent and supplementary to $\angle DCB$.

5. **ALGEBRA** Two angles are complementary. The measure of one angle is 21 more than twice the measure of the other angle. Find the measures of the angles.

6. **ALGEBRA** If a supplement of an angle has a measure 78 less than the measure of the angle, what are the measures of the angles?

ALGEBRA For Exercises 7–8, use the figure at the right.

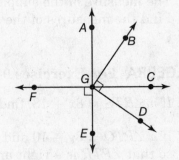

7. If $m\angle FGE = 5x + 10$, find the value of x so that $\overleftrightarrow{FC} \perp \overleftrightarrow{AE}$.

8. If $m\angle BGC = 16x - 4$ and $m\angle CGD = 2x + 13$, find the value of x so that $\angle BGD$ is a right angle.

Determine whether each statement can be assumed from the figure. Explain.

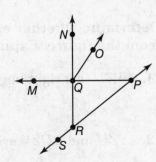

9. $\angle NQO$ and $\angle OQP$ are complementary.

10. $\angle SRQ$ and $\angle QRP$ is a linear pair.

11. $\angle MQN$ and $\angle MQR$ are vertical angles.

12. **STREET MAPS** Darren sketched a map of the cross streets nearest to his home for his friend Miguel. Describe two different angle relationships between the streets.

1-6 Skills Practice

Two-Dimensional Figures

Name each polygon by its number of sides and then classify it as *convex* or *concave* and *regular* or *irregular*.

1.

2.

3.

Find the perimeter or circumference of each figure. Round to the nearest tenth.

4.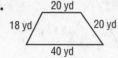

5.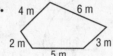

6.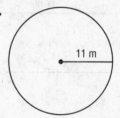

Find the area of each figure. Round to the nearest tenth.

7.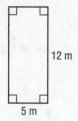

8.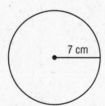

9.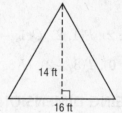

COORDINATE GEOMETRY Graph each figure with the given vertices and identify the figure. Find the perimeter and area of the figure.

10. $A(3, 5), B(3, 1), C(0, 1)$

11. $Q(-3, 2), R(1, 2), S(1, -4), T(-3, -4)$

12. $G(-4, 1), H(4, 1), I(0, -2)$

13. $K(-4, -2), L(-1, 2), M(8, 2), N(5, -2)$

1-6 Practice

Two-Dimensional Figures

Name each polygon by its number of sides and then classify it as *convex* or *concave* and *regular* or *irregular*.

1.

2.

3.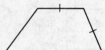

Find the perimeter or circumference and area of each figure. Round to the nearest tenth.

4.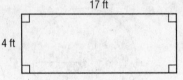
17 ft
4 ft

5.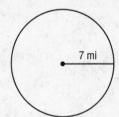
7 mi

6.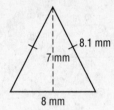
8.1 mm
7 mm
8 mm

COORDINATE GEOMETRY Graph each figure with the given vertices and identify the figure. Then find the perimeter and area of the figure.

7. $O(3, 2), P(1, 2), Q(1, -4), R(3, -4)$

8. $S(0, 0), T(3, -2), U(8, 0)$

CHANGING DIMENSIONS Use the rectangle from Exercise 4.

9. Suppose the length and width of the rectangle are doubled. What effect would this have on the perimeter? Justify your answer.

10. Suppose the length and width of the rectangle are doubled. What effect would this have on the area? Justify your answer.

11. **SEWING** Jasmine plans to sew fringe around the circular pillow shown in the diagram.

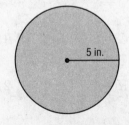
5 in.

   a. How many inches of fringe does she need to purchase?

   b. If Jasmine doubles the radius of the pillow, what is the new area of the top of the pillow?

1-7 Skills Practice

Three-Dimensional Figures

Determine if the solid is a polyhedron. Then identify the solid. If it is a polyhedron, name the bases, edges, and vertices.

1.

2.

3.

Find the surface area of each solid. Round to the nearest tenth.

4.

3 in.   6 in.

5.
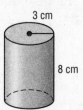
3 cm

8 cm

6.
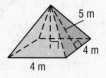
5 m

4 m   4 m

Find the volume of each solid. Round to the nearest tenth.

7.

6 ft
5 ft   4 ft

8.
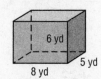
6 yd
8 yd   5 yd

9.
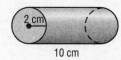
2 cm

10 cm

1-7 Practice

Three-Dimensional Figures

Determine whether the solid is a polyhedron. Then identify the solid. If it is a polyhedron, name the bases, edges, and vertices.

1.

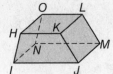

2.

Find the surface area and volume of each solid to the nearest tenth.

3.

4.

5.

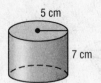

6. **COOKING** A cylindrical can of soup has a height of 4 inches and a radius of 2 inches. What is the volume of the can? Round to the nearest tenth.

7. **BUSINESS** A company needs boxes to hold a stack of 8.5 inch by 11 inch papers. If they would like the volume of the box to be 500 cubic inches, what should be the height of the box? Round to the nearest tenth.

2-1 Skills Practice

Inductive Reasoning and Conjecture

Write a conjecture that describes the pattern in the sequence. Then use your conjecture to find the next item in the sequence.

1.

2. $-4, -1, 2, 5, 8$ 3. $6, \frac{11}{2}, 5, \frac{9}{2}, 4$ 4. $-2, 4, -8, 16, -32$

Make a conjecture about each value or geometric relationship.

5. Points A, B, and C are collinear, and D is between B and C.

6. Point P is the midpoint of $\overline{NQ}$.

7. $\angle 1$, $\angle 2$, $\angle 3$, and $\angle 4$ form four linear pairs.

8. $\angle 3 \cong \angle 4$

Determine whether each conjecture is *true* or *false*. Give a counterexample for any false conjecture.

9. If $\angle ABC$ and $\angle CBD$ form a linear pair, then $\angle ABC \cong \angle CBD$.

10. If $\overline{AB}$, $\overline{BC}$, and $\overline{AC}$ are congruent, then A, B, and C are collinear.

11. If $AB + BC = AC$, then $AB = BC$.

12. If $\angle 1$ is complementary to $\angle 2$, and $\angle 1$ is complementary to $\angle 3$, then $\angle 2 \cong \angle 3$.

2-1 Practice

Inductive Reasoning and Conjecture

Make a conjecture about the next item in each sequence.

1.

2. $5, -10, 15, -20$

3. $-2, 1, -\frac{1}{2}, \frac{1}{4}, -\frac{1}{8}$

4. $12, 6, 3, 1.5, 0.75$

Make a conjecture about each value or geometric relationship.

5. $\angle ABC$ is a right angle.

6. Point S is between R and T.

7. $P, Q, R,$ and S are noncollinear and $\overline{PQ} \cong \overline{QR} \cong \overline{RS} \cong \overline{SP}$.

8. $ABCD$ is a parallelogram.

Determine whether each conjecture is *true* or *false*. Give a counterexample for any false conjecture.

9. If S, T, and U are collinear and $ST = TU$, then T is the midpoint of $\overline{SU}$.

10. If $\angle 1$ and $\angle 2$ are adjacent angles, then $\angle 1$ and $\angle 2$ form a linear pair.

11. If $\overline{GH}$ and $\overline{JK}$ form a right angle and intersect at P, then $\overline{GH} \perp \overline{JK}$.

12. **ALLERGIES** Each spring, Rachel starts sneezing when the pear trees on her street blossom. She reasons that she is allergic to pear trees. Find a counterexample to Rachel's conjecture.

2-2　Skills Practice

Logic

Use the following statements to write a compound statement for each conjunction or disjunction. Then find its truth value.

$p: -3 - 2 = -5$

$q:$ Vertical angles are congruent.

$r: 2 + 8 > 10$

$s:$ The sum of the measures of complementary angles is 90.

1. p and q

2. $p \wedge r$

3. p or s

4. $r \vee s$

5. $p \wedge \sim q$

6. $q \vee \sim r$

Complete each truth table.

7.

p	q	$\sim p$	$\sim p \wedge q$	$\sim(\sim p \wedge q)$
T	T			
T	F			
F	T			
F	F			

8.

p	q	$\sim q$	$\sim p \vee \sim q$
T	T	F	
T	F	T	
F	T	F	
F	F	T	

Construct a truth table for each compound statement.

9. $\sim q \wedge r$

10. $\sim p \vee \sim r$

2-2 Practice

Logic

Use the following statements to write a compound statement for each conjunction or disjunction. Then find its truth value.

p: 60 seconds = 1 minute
q: Congruent supplementary angles each have a measure of 90.
r: $-12 + 11 < -1$

1. $p \wedge q$

2. $q \vee r$

3. $\sim p \vee q$

4. $\sim p \wedge \sim r$

Complete each truth table.

5.

p	q	$\sim p$	$\sim q$	$\sim p \vee \sim q$
T	T			
T	F			
F	T			
F	F			

6.

p	q	$\sim p$	$\sim p \vee q$	$p \wedge (\sim p \vee q)$
T	T			
T	F			
F	T			
F	F			

Construct a truth table for each compound statement.

7. $q \vee (p \wedge \sim q)$

8. $\sim q \wedge (\sim p \vee q)$

9. SCHOOL The Venn diagram shows the number of students in the band who work after school or on the weekends.

a. How many students work after school and on weekends?

b. How many students work after school or on weekends?

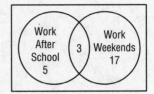

18   *Glencoe Geometry*

2-3 Skills Practice

Conditional Statements

Identify the hypothesis and conclusion of each conditional statement.

1. If you purchase a computer and do not like it, then you can return it within 30 days.

2. If $x + 8 = 4$, then $x = -4$.

3. If the drama class raises $2000, then they will go on tour.

Write each statement in if-then form.

4. A polygon with four sides is a quadrilateral.

5. An acute angle has a measure less than 90.

Determine the truth value of each conditional statement. If _true_, explain your reasoning. If _false_, give a counterexample.

6. If you have five dollars, then you have five one-dollar bills.

7. If I roll two six-sided dice and sum of the numbers is 11, then one die must be a five.

8. If two angles are supplementary, then one of the angles is acute.

9. Write the converse, inverse, and contrapositive of the conditional statement. Determine whether each statement is true or false. If a statement is false, find a counterexample.
If 89 is divisible by 2, then 89 is an even number.

2-3 Practice

Conditional Statements

Identify the hypothesis and conclusion of each conditional statement.

1. If $3x + 4 = -5$, then $x = -3$.

2. If you take a class in television broadcasting, then you will film a sporting event.

Write each statement in if-then form.

3. "Those who do not remember the past are condemned to repeat it." (*George Santayana*)

4. Adjacent angles share a common vertex and a common side.

Determine the truth value of each conditional statement. If *true*, explain your reasoning. If *false*, give a counterexample.

5. If a and b are negative, then $a + b$ is also negative.

6. If two triangles have equivalent angle measures, then they are congruent.

7. If the moon has purple spots, then it is June.

8. SUMMER CAMP Older campers who attend Woodland Falls Camp are expected to work. Campers who are juniors wait on tables.

   a. Write a conditional statement in if-then form.

   b. Write the converse of your conditional statement.

2-4 Skills Practice

Deductive Reasoning

Determine whether the stated conclusion is valid based on the given information. If not, write *invalid*. Explain your reasoning.

1. **Given:** If the sum of the measures of two angles is 180, then the angles are supplementary. $m\angle A + m\angle B$ is 180.

   Conclusion: $\angle A$ and $\angle B$ are supplementary.

2. **Given:** If the sum of the measures of two angles is 90, then the angles are complementary. $m\angle ABC$ is 45 and $m\angle DEF$ is 48.

   Conclusion: $\angle ABC$ and $\angle DEF$ are complementary.

3. **Given:** If the sum of the measures of two angles is 180, then the angles are supplementary. $\angle 1$ and $\angle 2$ are a linear pair.

   Conclusion: $\angle 1$ and $\angle 2$ are supplementary.

Use the Law of Syllogism to draw a valid conclusion from each set of statements, if possible. If no valid conclusion can be drawn write *no valid conclusion* and explain your reasoning.

4. If two angles are complementary, then the sum of their measures is 90.
   If the sum of the measures of two angles is 90, then both of the angles are acute.

5. If the heat wave continues, then air conditioning will be used more frequently.
   If air conditioning is used more frequently, then energy costs will be higher.

6. If it is Tuesday, then Marla tutors chemistry.
   If Marla tutors chemistry, then she arrives home at 4 P.M.

7. If a marine animal is a starfish, then it lives in the intertidal zone of the ocean.
   The intertidal zone is the least stable of the ocean zones.

2-4 Practice

Deductive Reasoning

Determine whether the stated conclusion is valid based on the given information. If not, write *invalid*. Explain your reasoning.

1. **Given:** If a point is the midpoint of a segment, then it divides the segment into two congruent segments. R is the midpoint of $\overline{QS}$

   Conclusion: $\overline{QR} \cong \overline{RS}$.

2. **Given:** If a point is the midpoint of a segment, then it divides the segment into two congruent segments. $\overline{AB} \cong \overline{BC}$

   Conclusion: B divides $\overline{AC}$ into two congruent segments.

Use the Law of Syllogism to draw a valid conclusion from each set of statements, if possible. If no valid conclusion can be drawn, write *no valid conclusion*.

3. If two angles form a linear pair, then the two angles are supplementary.
   If two angles are supplementary, then the sum of their measures is 180.

4. If a hurricane is Category 5, then winds are greater than 155 miles per hour.
   If winds are greater than 155 miles per hour, then trees, shrubs, and signs are blown down.

Draw a valid conclusion from the statements, if possible. Then state whether your conclusion was drawn using the Law of Detachment or the Law of Syllogism. If no valid conclusion can be drawn, write *no valid conclusion* and explain your reasoning.

5. **Given:** If a whole number is even, then its square is divisible by 4.
   The number I am thinking of is an even number.

6. **BIOLOGY** If an organism is a parasite, then it survives by living on or in a host organism. If a parasite lives in or on a host organism, then it harms its host. What conclusion can you draw if a virus is a parasite?

2-5 Skills Practice

Postulates and Paragraph Proofs

Explain how the figure illustrates that each statement is true. Then state the postulate that can be used to show each statement is true.

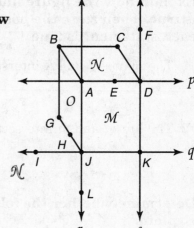

1. Planes O and M intersect in line r.

2. Line p lies in plane N.

Determine whether each statement is always, sometimes, or never true. Explain your reasoning.

3. Three collinear points determine a plane.

4. Two points A and B determine a line.

5. A plane contains at least three lines.

In the figure, $\overleftrightarrow{DG}$ and $\overrightarrow{DP}$ is in plane J and H lies on $\overleftrightarrow{DG}$. State the postulate that can be used to show each statement is true.

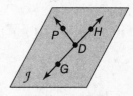

6. G and P are collinear.

7. Points D, H, and P are coplanar.

8. PROOF In the figure at the right, point B is the midpoint of $\overline{AC}$ and point C is the midpoint of $\overline{BD}$. Write a paragraph proof to prove that $AB = CD$.

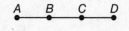

2-5 Practice

Postulates and Paragraph Proofs

Explain how the figure illustrates that each statement is true. Then state the postulate that can be used to show each statement is true.

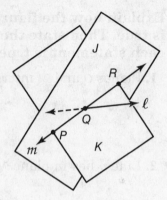

1. The planes J and K intersect at line m.

2. The lines l and m intersect at point Q.

Determine whether the following statements are *always*, *sometimes*, or *never* true. Explain.

3. The intersection of two planes contains at least two points.

4. If three planes have a point in common, then they have a whole line in common.

In the figure, line m and $\overrightarrow{TQ}$ lie in plane $\mathcal{A}$. State the postulate that can be used to show that each statement is true.

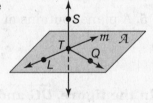

5. Points L, and T and line m lie in the same plane.

6. Line m and $\overleftrightarrow{ST}$ intersect at T.

7. In the figure, E is the midpoint of $\overline{AB}$ and $\overline{CD}$, and $AB = CD$. Write a paragraph proof to prove that $\overline{AE} \cong \overline{ED}$.

8. **LOGIC** Points A, B, and C are noncollinear. Points B, C, and D are noncollinear. Points A, B, C, and D are noncoplanar. Describe two planes that intersect in line BC.

2-6 Skills Practice

Algebraic Proof

State the property that justifies each statement.

1. If $80 = m\angle A$, then $m\angle A = 80$.

2. If $RS = TU$ and $TU = YP$, then $RS = YP$.

3. If $7x = 28$, then $x = 4$.

4. If $VR + TY = EN + TY$, then $VR = EN$.

5. If $m\angle 1 = 30$ and $m\angle 1 = m\angle 2$, then $m\angle 2 = 30$.

Complete the following proof.

6. **Given:** $8x - 5 = 2x + 1$
   Prove: $x = 1$
   Proof:

Statements	Reasons
a. $8x - 5 = 2x + 1$	**a.** _____
b. $8x - 5 - 2x = 2x + 1 - 2x$	**b.** _____
c. _____	**c.** Substitution Property
d. _____	**d.** Addition Property
e. $6x = 6$	**e.** _____
f. $\dfrac{6x}{6} = \dfrac{6}{6}$	**f.** _____
g. _____	**g.** _____

Write a two-column proof to verify the conjecture.

7. If $\overline{PQ} \cong \overline{QS}$ and $\overline{QS} \cong \overline{ST}$ then $PQ = ST$.

2-6 Practice

Algebraic Proof

PROOF Write a two-column proof to verify each conjecture.

1. If $m\angle ABC + m\angle CBD = 90$, $m\angle ABC = 3x - 5$,
   and $m\angle CBD = \dfrac{x + 1}{2}$, then $x = 27$.

2. **FINANCE** The formula for simple interest is $I = prt$, where I is interest, p is principal, r is rate, and t is time. Solve the formula for r and justify each step.

2-7 Skills Practice

Proving Segment Relationships

Justify each statement with a property of equality, a property of congruence, or a postulate.

1. $QA = QA$

2. If $\overline{AB} \cong \overline{BC}$ and $\overline{BC} \cong \overline{CE}$ then $\overline{AB} \cong \overline{CE}$.

3. If Q is between P and R, then $PR = PQ + QR$.

4. If $AB + BC = EF + FG$ and $AB + BC = AC$, then $EF + FG = AC$.

PROOF Complete each proof.

5. **Given:** $\overline{SU} \cong \overline{LR}$
   $\overline{TU} \cong \overline{LN}$
   Prove: $\overline{ST} \cong \overline{NR}$

Proof:

Statements	Reasons
a. $\overline{SU} \cong \overline{LR}, \overline{TU} \cong \overline{LN}$	a. _____
b. _____	b. Definition of $\cong$ segments
c. $SU = ST + TU$ $LR = LN + NR$	c. _____
d. $ST + TU = LN + NR$	d. _____
e. $ST + LN = LN + NR$	e. _____
f. $ST + LN - LN = LN + NR - LN$	f. _____
g. _____	g. Substitution Property
h. $\overline{ST} \cong \overline{NR}$	h. _____

6. **Given:** $\overline{AB} \cong \overline{CD}$
   Prove: $\overline{CD} \cong \overline{AB}$

Proof:

Statements	Reasons
a. _____	a. Given
b. $AB = CD$	b. _____
c. $CD = AB$	c. _____
d. _____	d. Definition of $\cong$ segments

2-7 Practice

Proving Segment Relationships

Complete the following proof.

1. **Given:** $\overline{AB} \cong \overline{DE}$
   B is the midpoint of $\overline{AC}$.
   E is the midpoint of $\overline{DF}$.

   Prove: $\overline{BC} \cong \overline{EF}$

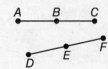

Proof:

Statements	Reasons
a._____ _____ _____	a. Given
b. $AB = DE$	b._____
c._____ _____	c. Definition of Midpoint
d. $BC = DE$	d._____
e. $BC = EF$	e._____
f._____	f._____

2. **TRAVEL** Refer to the figure. DeAnne knows that the distance from Grayson to Apex is the same as the distance from Redding to Pine Bluff. Prove that the distance from Grayson to Redding is equal to the distance from Apex to Pine Bluff.

Grayson   Apex   Redding   Pine Bluff
G   A   R   P

2-8 Skills Practice

Proving Angle Relationships

Find the measure of each numbered angle and name the theorems that justify your work.

1. $m\angle 2 = 57$

2. $m\angle 5 = 22$

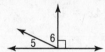

3. $m\angle 1 = 38$

4. $m\angle 13 = 4x + 11$,
   $m\angle 14 = 3x + 1$

5. $\angle 9$ and $\angle 10$ are
   complementary.
   $\angle 7 \cong \angle 9$, $m\angle 8 = 41$

6. $m\angle 2 = 4x - 26$,
   $m\angle 3 = 3x + 4$

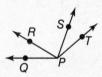

7. Complete the following proof.
   Given: $\angle QPS \cong \angle TPR$
   Prove: $\angle QPR \cong \angle TPS$
   Proof:

Statements	Reasons
a. _____	**a.** _____
b. $m\angle QPS = m\angle TPR$	**b.** _____
c. $m\angle QPS = m\angle QPR + m\angle RPS$ $m\angle TPR = m\angle TPS + m\angle RPS$	**c.** _____
d. _____	**d.** Substitution
_____	
e. _____	**e.** _____
f. _____	**f.** _____

2-8 Practice

Proving Angle Relationships

Find the measure of each numbered angle and name the theorems that justify your work.

1. $m\angle 1 = x + 10$
   $m\angle 2 = 3x + 18$

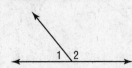

2. $m\angle 4 = 2x - 5$
   $m\angle 5 = 4x - 13$

3. $m\angle 6 = 7x - 24$
   $m\angle 7 = 5x + 14$

4. Write a two-column proof.

   Given: $\angle 1$ and $\angle 2$ form a linear pair.
   $\angle 2$ and $\angle 3$ are supplementary.

   Prove: $\angle 1 \cong \angle 3$

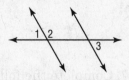

5. STREETS Refer to the figure. Barton Road and Olive Tree Lane form a right angle at their intersection. Tryon Street forms a 57° angle with Olive Tree Lane. What is the measure of the acute angle Tryon Street forms with Barton Road?

3-1 Skills Practice

Parallel Lines and Transversals

For Exercises 1–4, refer to the figure at the right to identify each of the following.

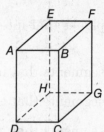

1. all planes that are parallel to plane *DEH*

2. all segments that are parallel to $\overline{AB}$

3. all segments that intersect $\overline{GH}$

4. all segments that are skew to $\overline{CD}$

Classify the relationship between each pair of angles as *alternate interior*, *alternate exterior*, *corresponding*, or *consecutive interior* angles.

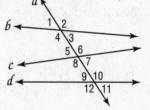

5. ∠4 and ∠5   **6.** ∠5 and ∠11

7. ∠4 and ∠6   **8.** ∠7 and ∠9

9. ∠2 and ∠8   **10.** ∠3 and ∠6

11. ∠1 and ∠9   **12.** ∠3 and ∠9

13. ∠6 and ∠12   **14.** ∠7 and ∠11

Identify the transversal connecting each pair of angles. Then classify the relationship between each pair of angles.

15. ∠4 and ∠10   **16.** ∠2 and ∠12

17. ∠7 and ∠3   **18.** ∠13 and ∠10

19. ∠8 and ∠14   **20.** ∠6 and ∠14

3-1 Practice

Parallel Lines and Transversals

Refer to the figure at the right to identify each of the following.

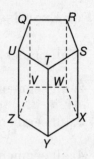

1. all planes that intersect plane *STX*

2. all segments that intersect $\overline{QU}$

3. all segments that are parallel to $\overline{XY}$

4. all segments that are skew to $\overline{VW}$

Classify the relationship between each pair of angles as *alternate interior*, *alternate exterior*, *corresponding*, or *consecutive interior* angles.

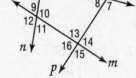

5. ∠2 and ∠10

6. ∠7 and ∠13

7. ∠9 and ∠13

8. ∠6 and ∠16

9. ∠3 and ∠10

10. ∠8 and ∠14

Name the transversal that forms each pair of angles. Then identify the special name for the angle pair.

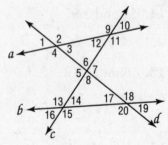

11. ∠2 and ∠12

12. ∠6 and ∠18

13. ∠13 and ∠19

14. ∠11 and ∠7

FURNITURE For Exercises 15–16, refer to the drawing of the end table.

15. Find an example of parallel planes.

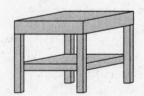

16. Find an example of parallel lines.

3-2 Skills Practice

Angles and Parallel Lines

In the figure, $m\angle 2 = 70$. Find the measure of each angle.

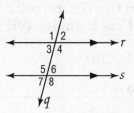

1. $\angle 3$ 2. $\angle 5$

3. $\angle 8$ 4. $\angle 1$

5. $\angle 4$ 6. $\angle 6$

In the figure, $m\angle 7 = 100$. Find the measure of each angle.

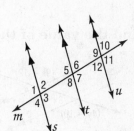

7. $\angle 9$ 8. $\angle 6$

9. $\angle 8$ 10. $\angle 2$

11. $\angle 5$ 12. $\angle 11$

In the figure, $m\angle 3 = 75$ and $m\angle 10 = 105$. Find the measure of each angle.

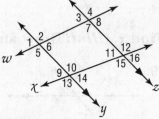

13. $\angle 2$ 14. $\angle 5$

15. $\angle 7$ 16. $\angle 15$

17. $\angle 14$ 18. $\angle 9$

Find the value of the variable(s) in each figure. Explain your reasoning.

19.
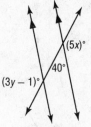

20.
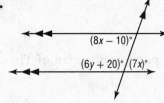

21.
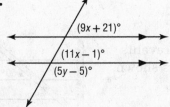

22.
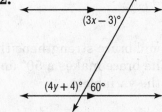

3-2 Practice

Angles and Parallel Lines

In the figure, $m\angle 2 = 92$ and $m\angle 12 = 74$. Find the measure of each angle. Tell which postulate(s) or theorem(s) you used.

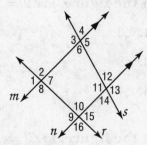

1. $\angle 10$ 2. $\angle 8$

3. $\angle 9$ 4. $\angle 5$

5. $\angle 11$ 6. $\angle 13$

Find the value of the variable(s) in each figure. Explain your reasoning.

7.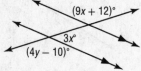

8.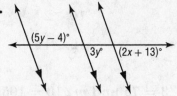

Find x. (Hint: Draw an auxiliary line.)

9.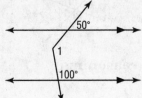

10.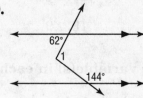

11. PROOF Write a paragraph proof of Theorem 3.3.

Given: $\ell \parallel m$, $m \parallel n$

Prove: $\angle 1 \cong \angle 12$

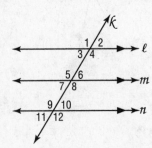

12. FENCING A diagonal brace strengthens the wire fence and prevents it from sagging. The brace makes a 50° angle with the wire as shown. Find the value of the variable.

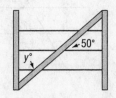

3-3 Skills Practice

Slopes of Lines

Determine the slope of the line that contains the given points.

1. $S(-1, 2)$, $W(0, 4)$

2. $G(-2, 5)$, $H(1, -7)$

3. $C(0, 1)$, $D(3, 3)$

4. $J(-5, -2)$, $K(5, -4)$

Find the slope of each line.

5.
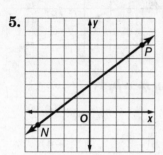

6.
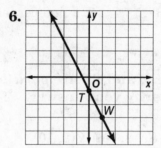

Determine whether $\overleftrightarrow{AB}$ and $\overleftrightarrow{MN}$ are *parallel*, *perpendicular*, or *neither*. Graph each line to verify your answer.

7. $A(0, 3)$, $B(5, -7)$, $M(-6, 7)$, $N(-2, -1)$

8. $A(-1, 4)$, $B(2, -5)$, $M(-3, 2)$, $N(3, 0)$

9. $A(-2, -7)$, $B(4, 2)$, $M(-2, 0)$, $N(2, 6)$

10. $A(-4, -8)$, $B(4, -6)$, $M(-3, 5)$, $N(-1, -3)$

Graph the line that satisfies each condition.

11. slope = 3, passes through $A(0, 1)$

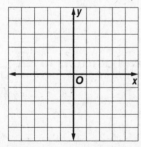

12. slope = $-\dfrac{3}{2}$, passes through $R(-4, 5)$

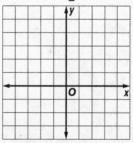

13. passes through $Y(3, 0)$, parallel to $\overleftrightarrow{DJ}$ with $D(-3, 1)$ and $J(3, 3)$

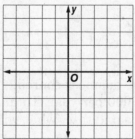

14. passes through $T(0, -2)$, perpendicular to $\overleftrightarrow{CX}$ with $C(0, 3)$ and $X(2, -1)$

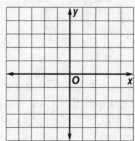

3-3 Practice

Slopes of Lines

Determine the slope of the line that contains the given points.

1. $B(-4, 4)$, $R(0, 2)$

2. $I(-2, -9)$, $P(2, 4)$

Find the slope of each line.

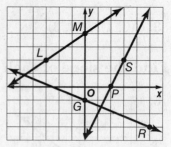

3. $\overleftrightarrow{LM}$

4. $\overleftrightarrow{GR}$

5. a line parallel to $\overleftrightarrow{GR}$

6. a line perpendicular to $\overrightarrow{PS}$

Determine whether $\overleftrightarrow{KM}$ and $\overleftrightarrow{ST}$ are *parallel*, *perpendicular*, or *neither*. Graph each line to verify your answer.

7. $K(-1, -8)$, $M(1, 6)$, $S(-2, -6)$, $T(2, 10)$

8. $K(-5, -2)$, $M(5, 4)$, $S(-3, 6)$, $T(3, -4)$

9. $K(-4, 10)$, $M(2, -8)$, $S(1, 2)$, $T(4, -7)$

10. $K(-3, -7)$, $M(3, -3)$, $S(0, 4)$, $T(6, -5)$

Graph the line that satisfies each condition.

11. slope $= -\dfrac{1}{2}$, contains $U(2, -2)$

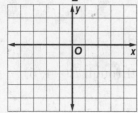

12. slope $= \dfrac{4}{3}$, contains $P(-3, -3)$

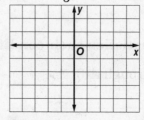

13. contains $B(-4, 2)$, parallel to $\overleftrightarrow{FG}$ with $F(0, -3)$ and $G(4, -2)$

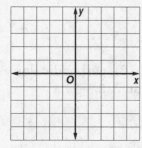

14. contains $Z(-3, 0)$, perpendicular to $\overleftrightarrow{EK}$ with $E(-2, 4)$ and $K(2, -2)$

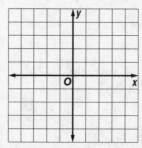

15. PROFITS After Take Two began renting DVDs at their video store, business soared. Between 2005 and 2010, profits increased at an average rate of $9000 per year. Total profits in 2010 were $45,000. If profits continue to increase at the same rate, what will the total profit be in 2014?

3-4 Skills Practice

Equations of Lines

Write an equation in slope-intercept form of the line having the given slope and y-intercept. Then graph the line.

1. m: -4, b: 3

2. m: 3, b: -8

3. m: $\frac{3}{7}$, $(0, 1)$

4. m: $-\frac{2}{5}$, $(0, -6)$

Write equations in point-slope form of the line having the given slope that contains the given point. Then graph the line.

5. $m = 2$, $(5, 2)$

6. $m = -3$, $(2, -4)$

7. $m = -\frac{1}{2}$, $(-2, 5)$

8. $m = \frac{1}{3}$, $(-3, -8)$

Write an equation in slope-intercept form for each line shown or described.

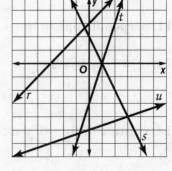

9. r

10. s

11. t

12. u

13. the line parallel to line r that contains $(1, -1)$

14. the line perpendicular to line s that contains $(0, 0)$

15. $m = 6$, $b = -2$

16. $m = -\frac{5}{3}$, $b = 0$

17. $m = -1$, contains $(0, -6)$

18. $m = 4$, contains $(2, 5)$

19. contains $(2, 0)$ and $(0, 10)$

20. x-intercept is -2, y-intercept is -1

3-4 Practice

Equations of Lines

Write an equation in slope-intercept form of the line having the given slope and y-intercept or given points. Then graph the line.

1. $m: \frac{2}{3}$, $b: -10$　　　　**2.** $m: -\frac{7}{9}$, $\left(0, -\frac{1}{2}\right)$　　　　**3.** $m: 4.5$, $(0, 0.25)$

Write equations in point-slope form of the line having the given slope that contains the given point. Then graph the line.

4. $m: \frac{3}{2}$, $(4, 6)$　　　　　　　　　　**5.** $m: -\frac{6}{5}$, $(-5, -2)$

6. $m: 0.5$, $(7, -3)$　　　　　　　　　　**7.** $m: -1.3$, $(-4, 4)$

Write an equation in slope-intercept form for each line shown or described.

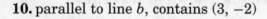

8. b　　　　　　　　　　**9.** c

10. parallel to line b, contains $(3, -2)$

11. perpendicular to line c, contains $(-2, -4)$

12. $m = -\frac{4}{9}$, $b = 2$　　　　　　**13.** $m = 3$, contains $(2, -3)$

14. x-intercept is -6, y-intercept is 2　　　　**15.** x-intercept is 2, y-intercept is -5

16. passes through $(2, -4)$ and $(5, 8)$　　　　**17.** contains $(-4, 2)$ and $(8, -1)$

18. COMMUNITY EDUCATION A local community center offers self-defense classes for teens. A \$25 enrollment fee covers supplies and materials and open classes cost \$10 each. Write an equation to represent the total cost of x self-defense classes at the community center.

3-5 Skills Practice

Proving Lines Parallel

Given the following information, determine which lines, if any, are parallel. State the postulate or theorem that justifies your answer.

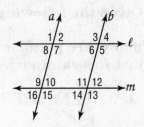

1. $\angle 3 \cong \angle 7$

2. $\angle 9 \cong \angle 11$

3. $\angle 2 \cong \angle 16$

4. $m\angle 5 + m\angle 12 = 180$

Find x so that $\ell \parallel m$. Show your work.

5.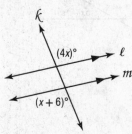

6.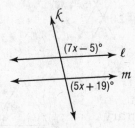

7.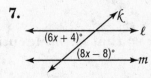

8.

9.

10.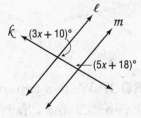

11. PROOF Provide a reason for each statement in the proof of Theorem 3.7.

Given: $\angle 1$ and $\angle 2$ are complementary.
$\overline{BC} \perp \overline{CD}$

Prove: $\overline{BA} \parallel \overline{CD}$

Proof:

Statements	Reasons
1. $\overline{BC} \perp \overline{CD}$	1.
2. $m\angle ABC = m\angle 1 + m\angle 2$	2.
3. $\angle 1$ and $\angle 2$ are complementary.	3.
4. $m\angle 1 + m\angle 2 = 90$	4.
5. $m\angle ABC = 90$	5.
6. $\overline{BA} \perp \overline{BC}$	6.
7. $\overline{BA} \parallel \overline{CD}$	7.

3-5   Practice

Proving Lines Parallel

Given the following information, determine which lines,

if any, are parallel. State the postulate or theorem that
justifies your answer.

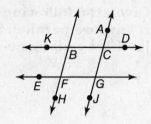

1. $m\angle BCG + m\angle FGC = 180$ **2.** $\angle CBF \cong \angle GFH$

3. $\angle EFB \cong \angle FBC$ **4.** $\angle ACD \cong \angle KBF$

Find x so that $l \parallel m$. Identify the postulate or theorem you used.

5.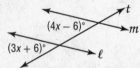

6.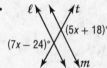

7.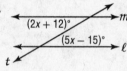

8. PROOF Write a two-column proof.

   Given: $\angle 2$ and $\angle 3$ are supplementary.
   Prove: $\overline{AB} \parallel \overline{CD}$

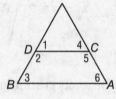

9. LANDSCAPING The head gardener at a botanical garden wants to plant rosebushes in
parallel rows on either side of an existing footpath. How can the gardener ensure that
the rows are parallel?

3-6 Skills Practice

Perpendiculars and Distance

Construct the segment that represents the distance indicated.

1. B to $\overleftrightarrow{AC}$

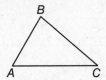

2. G to $\overleftrightarrow{EF}$

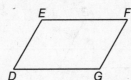

3. Q to $\overleftrightarrow{SR}$

COORDINATE GEOMETRY Find the distance from P to ℓ.

4. Line ℓ contains points $(0, -2)$ and $(6, 6)$. Point P has coordinates $(-1, 5)$.

5. Line ℓ contains points $(2, 4)$ and $(5, 1)$. Point P has coordinates $(1, 1)$.

6. Line ℓ contains points $(-4, -2)$ and $(2, 0)$. Point P has coordinates $(3, 7)$.

7. Line ℓ contains points $(-7, 8)$ and $(0, 5)$. Point P has coordinates $(-5, 32)$.

Find the distance between each pair of parallel lines with the given equations.

8. $y = 7$
$y = -1$

9. $x = -6$
$x = 5$

10. $y = 3x$
$y = 3x + 10$

11. $y = -5x$
$y = -5x + 26$

12. $y = x + 9$
$y = x + 3$

13. $y = -2x + 5$
$y = -2x - 5$

3-6 Practice

Perpendiculars and Distance

Construct the segment that represents the distance indicated.

1. O to $\overleftrightarrow{MN}$

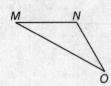

2. A to $\overleftrightarrow{DC}$

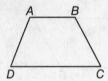

3. T to $\overleftrightarrow{VU}$

COORDINATE GEOMETRY Find the distance from P to ℓ.

4. Line ℓ contains points $(-2, 0)$ and $(4, 8)$. Point P has coordinates $(5, 1)$.

5. Line ℓ contains points $(3, 5)$ and $(7, 9)$. Point P has coordinates $(2, 10)$.

6. Line ℓ contains points $(5, 18)$ and $(9, 10)$. Point P has coordinates $(-4, 26)$.

7. Line ℓ contains points $(-2, 4)$ and $(1, -9)$. Point P has coordinates $(14, -6)$.

Find the distance between each pair of parallel lines with the given equation.

8. $y = -x$
   $y = -x - 4$

9. $y = 2x + 7$
   $y = 2x - 3$

10. $y = 3x + 12$
    $y = 3x - 18$

11. Graph the line $y = -x + 1$. Construct a perpendicular segment through the point at $(-2, -3)$. Then find the distance from the point to the line.

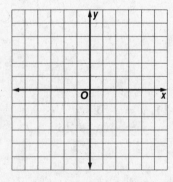

12. **CANOEING** Bronson and a friend are going to carry a canoe across a flat field to the bank of a straight canal. Describe the shortest path they can use.

4-1 Skills Practice

Classifying Triangles

Classify each triangle as *acute, equiangular, obtuse,* or *right*.

1.

2.

3.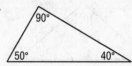

4.

5.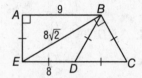

6.

Classify each triangle as *equilateral, isosceles,* or *scalene*.

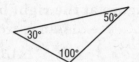

7. △*ABE*

8. △*EDB*

9. △*EBC*

10. △*DBC*

11. **ALGEBRA** Find *x* and the length of each side if △*ABC* is an isosceles triangle with $\overline{AB} \cong \overline{BC}$.

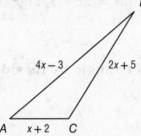

12. **ALGEBRA** Find *x* and the length of each side if △*FGH* is an equilateral triangle.

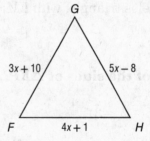

13. **ALGEBRA** Find *x* and the length of each side if △*RST* is an isosceles triangle with $\overline{RS} \cong \overline{TS}$.

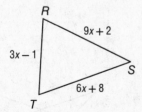

14. **ALGEBRA** Find *x* and the length of each side if △*DEF* is an equilateral triangle.

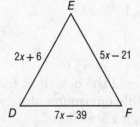

4-1 Practice

Classifying Triangles

Classify each triangle as *acute, equiangular, obtuse*, or *right*.

1.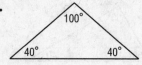

2.

3.

Classify each triangle in the figure at the right by its angles and sides.

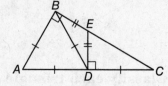

4. △ABD

5. △ABC

6. △EDC

7. △BDC

ALGEBRA For each triangle, find x and the measure of each side.

8. △FGH is an equilateral triangle with $FG = x + 5$, $GH = 3x - 9$, and $FH = 2x - 2$.

9. △LMN is an isosceles triangle, with $LM = LN$, $LM = 3x - 2$, $LN = 2x + 1$, and $MN = 5x - 2$.

Find the measures of the sides of △KPL and classify each triangle by its sides.

10. $K(-3, 2)$, $P(2, 1)$, $L(-2, -3)$

11. $K(5, -3)$, $P(3, 4)$, $L(-1, 1)$

12. $K(-2, -6)$, $P(-4, 0)$, $L(3, -1)$

13. **DESIGN** Diana entered the design at the right in a logo contest
sponsored by a wildlife environmental group. Use a protractor.
How many right angles are there?

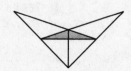

4-2 Skills Practice

Angles of Triangles

Find the measure of each numbered angle.

1.

2.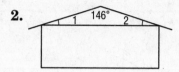

Find each measure.

3. $m\angle 1$

4. $m\angle 2$

5. $m\angle 3$

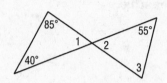

Find each measure.

6. $m\angle 1$

7. $m\angle 2$

8. $m\angle 3$

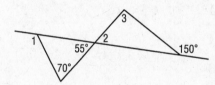

Find each measure.

9. $m\angle 1$

10. $m\angle 2$

11. $m\angle 3$

12. $m\angle 4$

13. $m\angle 5$

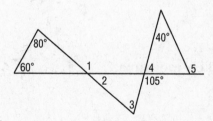

Find each measure.

14. $m\angle 1$

15. $m\angle 2$

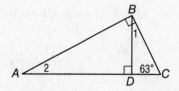

4-2 Practice

Angles of Triangles

Find the measure of each numbered angle.

1.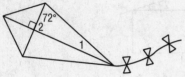

2.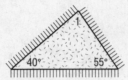

Find each measure.

3. $m\angle 1$

4. $m\angle 2$

5. $m\angle 3$

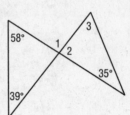

Find each measure.

6. $m\angle 1$

7. $m\angle 4$

8. $m\angle 3$

9. $m\angle 2$

10. $m\angle 5$

11. $m\angle 6$

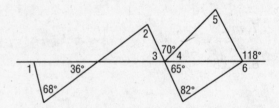

Find each measure.

12. $m\angle 1$

13. $m\angle 2$

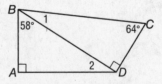

14. CONSTRUCTION The diagram shows an example of the Pratt Truss used in bridge construction. Use the diagram to find $m\angle 1$.

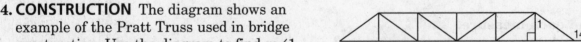

4-3 Skills Practice

Congruent Triangles

Show that polygons are congruent by identifying all congruent corresponding parts. Then write a congruence statement.

1.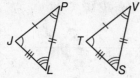

2.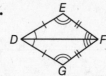

In the figure, $\triangle ABC \cong \triangle FDE$.

3. Find the value of x.

4. Find the value of y.

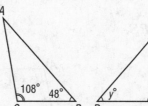

5. **PROOF** Write a two-column proof.

    Given: $\overline{AB} \cong \overline{CB}, \overline{AD} \cong \overline{CD}$, $\angle ABD \cong \angle CBD$,
            $\angle ADB \cong \angle CDB$

    Prove: $\triangle ABD \cong \triangle CBD$

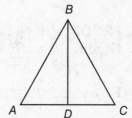

4-3 Practice

Congruent Triangles

Show that the polygons are congruent by indentifying all congruent corresponding parts. Then write a congruence statement.

1.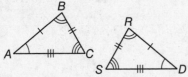

2.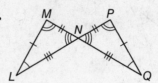

Polygon $ABCD \cong$ polygon $PQRS$.

3. Find the value of x.

4. Find the value of y.

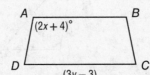

5. **PROOF** Write a two-column proof.
   Given: $\angle P \cong \angle R$, $\angle PSQ \cong \angle RSQ$, $\overline{PQ} \cong \overline{RQ}$, $\overline{PS} \cong \overline{RS}$

   Prove: $\triangle PQS \cong \triangle RQS$

6. **QUILTING**

   a. Indicate the triangles that appear to be congruent.

   b. Name the congruent angles and congruent sides of a pair of congruent triangles.

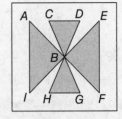

4-4   Skills Practice

Proving Triangles Congruent—SSS, SAS

Determine whether $\triangle ABC \cong \triangle KLM$. **Explain.**

1. $A(-3, 3)$, $B(-1, 3)$, $C(-3, 1)$, $K(1, 4)$, $L(3, 4)$, $M(1, 6)$

2. $A(-4, -2)$, $B(-4, 1)$, $C(-1, -1)$, $K(0, -2)$, $L(0, 1)$, $M(4, 1)$

PROOF Write the specified type of proof.

3. Write a flow proof.
   Given: $\overline{PR} \cong \overline{DE}$, $\overline{PT} \cong \overline{DF}$
   $\angle R \cong \angle E$, $\angle T \cong \angle F$
   Prove: $\triangle PRT \cong \triangle DEF$

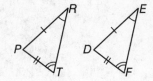

4. Write a two-column proof.
   Given: $\overline{AB} \cong \overline{CB}$, D is the midpoint of $\overline{AC}$.
   Prove: $\triangle ABD \cong \triangle CBD$

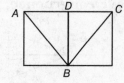

4-4 Practice

Proving Triangles Congruent—SSS, SAS

Determine whether $\triangle DEF \cong \triangle PQR$ given the coordinates of the vertices. Explain.

1. $D(-6, 1), E(1, 2), F(-1, -4), P(0, 5), Q(7, 6), R(5, 0)$

2. $D(-7, -3), E(-4, -1), F(-2, -5), P(2, -2), Q(5, -4), R(0, -5)$

3. Write a flow proof.
   Given: $\overline{RS} \cong \overline{TS}$
          V is the midpoint of $\overline{RT}$.
   Prove: $\triangle RSV \cong \triangle TSV$

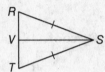

Determine which postulate can be used to prove that the triangles are congruent. If it is not possible to prove congruence, write *not possible*.

4.

5.

6.
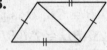

7. **INDIRECT MEASUREMENT** To measure the width of a sinkhole on his property, Harmon marked off congruent triangles as shown in the diagram. How does he know that the lengths $A'B'$ and AB are equal?

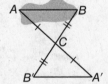

             50

4-5 Skills Practice

Proving Triangles Congruent—ASA, AAS

PROOF Write a flow proof.

1. Given: $\angle N \cong \angle L$

   $\overline{JK} \cong \overline{MK}$

   Prove: $\triangle JKN \cong \triangle MKL$

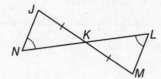

2. Given: $\overline{AB} \cong \overline{CB}$

   $\angle A \cong \angle C$

   $\overline{DB}$ bisects $\angle ABC$.

   Prove: $\overline{AD} \cong \overline{CD}$

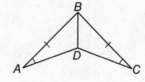

3. Write a paragraph proof.

   Given: $\overline{DE} \parallel \overline{FG}$

   $\angle E \cong \angle G$

   Prove: $\triangle DFG \cong \triangle FDE$

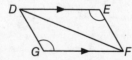

4-5 Practice

Proving Triangles Congruent—ASA, AAS

PROOF Write the specified type of proof.

1. Write a flow proof.

   Given: S is the midpoint of $\overline{QT}$.
   $\overline{QR} \parallel \overline{TU}$

   Prove: $\triangle QSR \cong \triangle TSU$

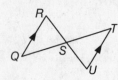

2. Write a paragraph proof.

   Given: $\angle D \cong \angle F$
   $\overline{GE}$ bisects $\angle DEF$.

   Prove: $\overline{DG} \cong \overline{FG}$

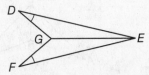

ARCHITECTURE For Exercises 3 and 4, use the following information.

An architect used the window design in the diagram when remodeling an art studio. $\overline{AB}$ and $\overline{CB}$ each measure 3 feet.

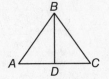

3. Suppose D is the midpoint of $\overline{AC}$. Determine whether $\triangle ABD \cong \triangle CBD$. Justify your answer.

4. Suppose $\angle A \cong \angle C$. Determine whether $\triangle ABD \cong \triangle CBD$. Justify your answer.

4-6 Skills Practice

Isosceles and Equilateral Triangles

Refer to the figure at the right.

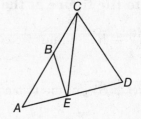

1. If $\overline{AC} \cong \overline{AD}$, name two congruent angles.

2. If $\overline{BE} \cong \overline{BC}$, name two congruent angles.

3. If $\angle EBA \cong \angle EAB$, name two congruent segments.

4. If $\angle CED \cong \angle CDE$, name two congruent segments.

Find each measure.

5. $m\angle ABC$

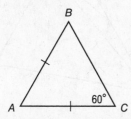

6. $m\angle EDF$

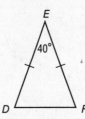

ALGEBRA Find the value of each variable.

7.

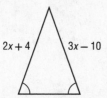

8.

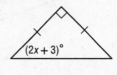

9. **PROOF** Write a two-column proof.

    Given: $\overline{CD} \cong \overline{CG}$

            $\overline{DE} \cong \overline{GF}$

    Prove: $\overline{CE} \cong \overline{CF}$

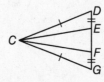

4-6   **Practice**

Isosceles and Equilateral Triangles

Refer to the figure at the right.

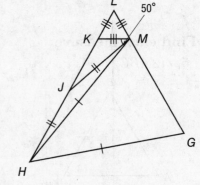

1. If $\overline{RV} \cong \overline{RT}$, name two congruent angles.

2. If $\overline{RS} \cong \overline{SV}$, name two congruent angles.

3. If $\angle SRT \cong \angle STR$, name two congruent segments.

4. If $\angle STV \cong \angle SVT$, name two congruent segments.

Find each measure.

5. $m\angle KML$ 6. $m\angle HMG$ 7. $m\angle GHM$

8. If $m\angle HJM = 145$, find $m\angle MHJ$.

9. If $m\angle G = 67$, find $m\angle GHM$.

10. **PROOF** Write a two-column proof.

    Given: $\overline{DE} \parallel \overline{BC}$
            $\angle 1 \cong \angle 2$
    Prove: $\overline{AB} \cong \overline{AC}$

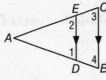

11. **SPORTS** A pennant for the sports teams at Lincoln High
School is in the shape of an isosceles triangle. If the measure
of the vertex angle is 18°, find the measure of each base angle.

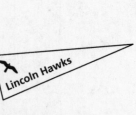

4-7 Skills Practice

Congruence Transformations

Identify the type of congruence transformation shown as a *reflection*, *translation*, or *rotation*.

1.
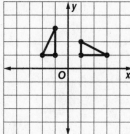

2.
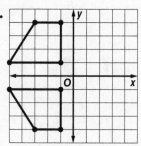

COORDINATE GEOMETRY Identify each transformation and verify that it is a congruence transformation.

3.
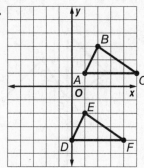

4.
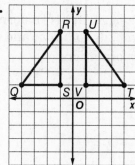

COORDINATE GEOMETRY Graph each pair of triangles with the given vertices. Then, identify the transformation, and verify that it is a congruence transformation.

5. $A(1, 3)$, $B(1, 1)$, $C(4, 1)$;
   $D(3, -1)$, $E(1, -1)$, $F(1, -4)$

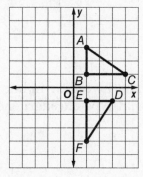

6. $J(-3, 0)$, $K(-2, 4)$, $L(-1, 0)$;
   $Q(2, -4)$, $R(3, 0)$, $S(4, -4)$

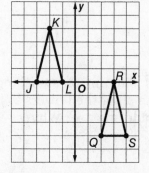

4-7 Practice

Congruence Transformations

Identify the type of congruence transformation shown as a *reflection*, *translation*, or *rotation*.

1.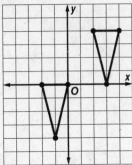

2.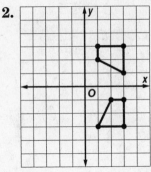

3. Identify the type of congruence transformation shown as a *reflection*, *translation*, or *rotation*, and verify that it is a congruence transformation.

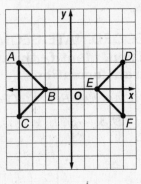

4. $\triangle ABC$ has vertices $A(-4, 2)$, $B(-2, 0)$, $C(-4, -2)$. $\triangle DEF$ has vertices $D(4, 2)$, $E(2, 0)$, $F(4, -2)$. Graph the original figure and its image. Then identify the transformation and verify that it is a congruence transformation.

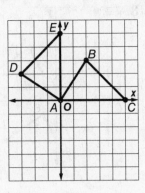

5. STENCILS Carly is planning on stenciling a pattern of flowers along the ceiling in her bedroom. She wants all of the flowers to look exactly the same. What type of congruence transformation should she use? Why?

4-8 Skills Practice

Triangles and Coordinate Proof

Position and label each triangle on the coordinate plane.

1. right △*FGH* with legs *a* units and *b* units long

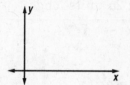

2. isosceles △*KLP* with base $\overline{KP}$ 6*b* units long

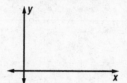

3. isosceles △*AND* with base $\overline{AD}$ 5*a* units long

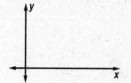

Name the missing coordinates of each triangle.

4.

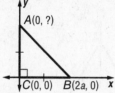

5.

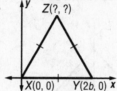

6.

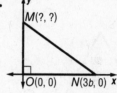

7.

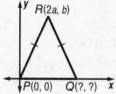

8.

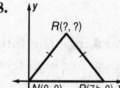

9.

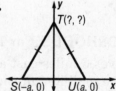

10. **PROOF** Write a coordinate proof to prove that in an isosceles right triangle, the segment from the vertex of the right angle to the midpoint of the hypotenuse is perpendicular to the hypotenuse.

Given: Isosceles right △*ABC* with right ∠*ABC*; *M* is the midpoint of $\overline{AC}$.

Prove: $\overline{BM} \perp \overline{AC}$

4-8 Practice

Triangles and Coordinate Proof

Position and label each triangle on the coordinate plane.

1. equilateral $\triangle SWY$ with sides $\frac{1}{4}a$ units long

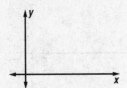

2. isosceles $\triangle BLP$ with base $\overline{BL}$ $3b$ units long

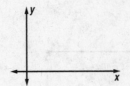

3. isosceles right $\triangle DGJ$ with hypotenuse $\overline{DJ}$ and legs $2a$ units long

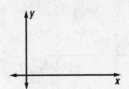

Name the missing coordinates of each triangle.

4.
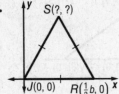
$S(?, ?)$
$J(0, 0)$ $R\left(\frac{1}{3}b, 0\right)$

5.
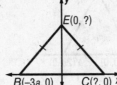
$E(0, ?)$
$B(-3a, 0)$ $C(?, 0)$

6.

$M(0, ?)$
$N(?, 0)$ $P(2b, 0)$

NEIGHBORHOODS For Exercises 7 and 8, use the following information.

Karina lives 6 miles east and 4 miles north of her high school. After school she works part time at the mall in a music store. The mall is 2 miles west and 3 miles north of the school.

7. **Proof** Write a coordinate proof to prove that Karina's high school, her home, and the mall are at the vertices of a right triangle.

   Given: $\triangle SKM$
   Prove: $\triangle SKM$ is a right triangle.

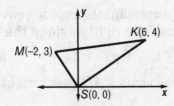

$M(-2, 3)$ $K(6, 4)$ $S(0, 0)$

8. Find the distance between the mall and Karina's home.

5-1 Skills Practice

Bisectors of Triangles

Find each measure.

1. *FG*

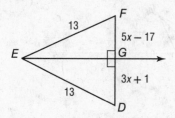

2. *KL*

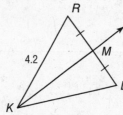

3. *TU*

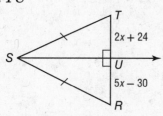

4. ∠*LYF*

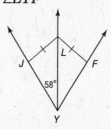

5. *IU*

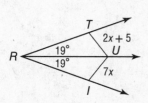

6. ∠*MYW*

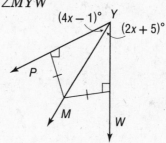

Point *P* is the circumcenter of △*ABC*. List any segment(s) congruent to each segment.

7. $\overline{BR}$

8. $\overline{CS}$

9. $\overline{BP}$

Point *A* is the incenter of △*PQR*. Find each measure.

10. ∠*ARU*

11. *AU*

12. ∠*QPK*

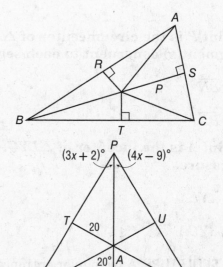

5-1 Practice

Bisectors of Triangles

Find each measure.

1. *TP*

2. *VU*

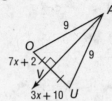

3. *KN*

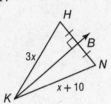

4. ∠*NJZ*

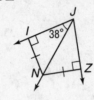

5. *QA*

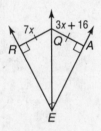

6. ∠*MFZ*

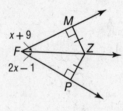

Point P is the circumcenter of △ABC. List any segment(s) congruent to each segment.

7. $\overline{BN}$

8. $\overline{BL}$

Point A is the incenter of △LYG. Find each measure.

9. ∠*YLA*

10. ∠*YGA*

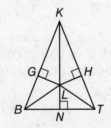

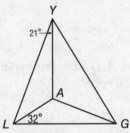

11. **SCULPTURE** A triangular entranceway has walls with corner angles of 50°, 70°, and 60°. The designer wants to place a tall bronze sculpture on a round pedestal in a central location equidistant from the three walls. How can the designer find where to place the sculpture?

5-2   Skills Practice

Medians and Altitudes of Triangles

In △PQR, NQ = 6, RK = 3, and PK = 4.
Find each measure.

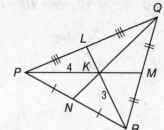

1. KM

2. KQ

3. LK

4. LR

5. NK

6. PM

In △STR, H is the centroid, EH = 6,
DH = 4, and SM = 24. Find each measure.

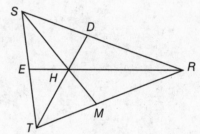

7. SH

8. HM

9. TH

10. HR

11. TD

12. ER

COORDINATE GEOMETRY Find the coordinates of the centroid of the triangle with
the given vertices.

13. X(–3, 15) Y(1, 5), Z(5, 10)

14. S(2, 5), T(6, 5), R(10, 0)

COORDINATE GEOMETRY Find the coordinates of the orthocenter of the triangle
with the given vertices.

15. L(8, 0), M(10, 8), N(14, 0)

16. D(–9, 9), E(–6, 6), F(0, 6)

5-2 Practice

Medians and Altitudes of Triangles

In $\triangle ABC$, $CP = 30$, $EP = 18$, and $BF = 39$. Find each measure.

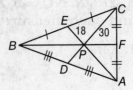

1. PD

2. FP

3. BP

4. CD

5. PA

6. EA

In $\triangle MIV$, Z is the centroid, $MZ = 6$, $YI = 18$, and $NZ = 12$. Find each measure.

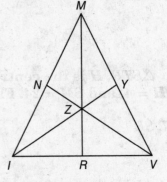

7. ZR

8. YZ

9. MR

10. ZV

11. NV

12. IZ

COORDINATE GEOMETRY Find the coordinates of the centroid of the triangle with the given vertices.

13. $I(3, 1)$, $J(6, 3)$, $K(3, 5)$

14. $H(0, 1)$, $U(4, 3)$, $P(2, 5)$

COORDINATE GEOMETRY Find the coordinates of the orthocenter of the triangle with the given vertices.

15. $P(-1, 2)$, $Q(5, 2)$, $R(2, 1)$

16. $S(0, 0)$, $T(3, 3)$, $U(3, 6)$

17. **MOBILES** Nabuko wants to construct a mobile out of flat triangles so that the surfaces of the triangles hang parallel to the floor when the mobile is suspended. How can Nabuko be certain that she hangs the triangles to achieve this effect?

5-3 Skills Practice

Inequalities in One Triangle

Use the Exterior Angle Inequality Theorem to list all of the angles that satisfy the stated condition.

1. measures less than $m\angle 1$

2. measures less than $m\angle 9$

3. measures greater than $m\angle 5$

4. measures greater than $m\angle 8$

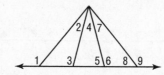

List the angles and sides of each triangle in order from smallest to largest.

5.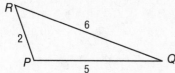

6.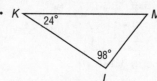

7.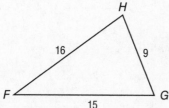

8.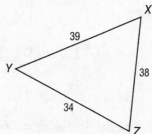

9.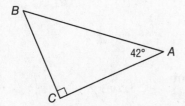

10.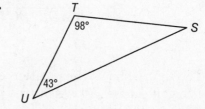

5-3 Practice

Inequalities in One Triangle

Use the figure at the right to determine which angle has the greatest measure.

1. ∠1, ∠3, ∠4

2. ∠4, ∠8, ∠9

3. ∠2, ∠3, ∠7

4. ∠7, ∠8, ∠10

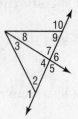

Use the Exterior Angle Inequality Theorem to list all angles that satisfy the stated condition.

5. measures are less than m∠1

6. measures are less than m∠3

7. measures are greater than m∠7

8. measures are greater than m∠2

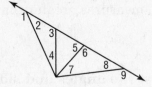

Use the figure at the right to determine the relationship between the measures of the given angles.

9. m∠QRW, m∠RWQ

10. m∠RTW, m∠TWR

11. m∠RST, m∠TRS

12. m∠WQR, m∠QRW

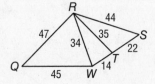

Use the figure at the right to determine the relationship between the lengths of the given sides.

13. $\overline{DH}$, $\overline{GH}$

14. $\overline{DE}$, $\overline{DG}$

15. $\overline{EG}$, $\overline{FG}$

16. $\overline{DE}$, $\overline{EG}$

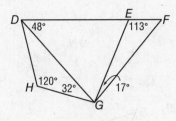

17. **SPORTS** The figure shows the position of three trees on one part of a disc golf course. At which tree position is the angle between the trees the greatest?

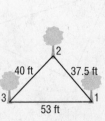

5-4 **Skills Practice**

Indirect Proof

State the assumption you would make to start an indirect proof of each statement.

1. $m\angle ABC < m\angle CBA$

2. $\triangle DEF \cong \triangle RST$

3. Line a is perpendicular to line b.

4. $\angle 5$ is supplementary to $\angle 6$.

Write an indirect proof of each statement.

5. **Given:** $x^2 + 8 \leq 12$
   Prove: $x \leq 2$

6. **Given:** $\angle D \not\cong \angle F$
   Prove: $DE \neq EF$

5-4 Practice

Indirect Proof

State the assumption you would make to start an indirect proof of each statement.

1. $\overline{BD}$ bisects $\angle ABC$.

2. $RT = TS$

Write an indirect proof of each statement.

3. **Given:** $-4x + 2 < -10$
   Prove: $x > 3$

4. **Given:** $m\angle 2 + m\angle 3 \neq 180$
   Prove: $a \nparallel b$

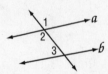

5. **PHYSICS** Sound travels through air at about 344 meters per second when the temperature is 20°C. If Enrique lives 2 kilometers from the fire station and it takes 5 seconds for the sound of the fire station siren to reach him, how can you prove indirectly that it is not 20°C when Enrique hears the siren?

5-5 Skills Practice

The Triangle Inequality

Is it possible to form a triangle with the given side lengths? If not, explain why not.

1. 2 ft, 3 ft, 4 ft

2. 5 m, 7 m, 9 m

3. 4 mm, 8 mm, 11 mm

4. 13 in., 13 in., 26 in.

5. 9 cm, 10 cm, 20 cm

6. 15 km, 17 km, 19 km

7. 14 yd, 17 yd, 31 yd

8. 6 m, 7 m, 12 m

Find the range for the measure of the third side of a triangle given the measures of two sides.

9. 5 ft, 9 ft

10. 7 in., 14 in.

11. 8 m, 13 m

12. 10 mm, 12 mm

13. 12 yd, 15 yd

14. 15 km, 27 km

15. 17 cm, 28 cm,

16. 18 ft, 22 ft

17. Proof Complete the proof.

Given: $\triangle ABC$ and $\triangle CDE$

Prove: $AB + BC + CD + DE > AE$

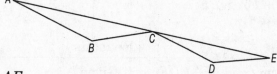

Proof:

Statements	Reasons
1. $AB + BC > AC$ $CD + DE > CE$	1. _____
2. $AB + BC + CD + DE > AC + CE$	2. _____
3. _____	3. Seg. Addition Post
4. _____	4. Substitution

5-5 Practice

The Triangle Inequality

Is it possible to form a triangle with the given side lengths? If not explain why not.

1. 9, 12, 18

2. 8, 9, 17

3. 14, 14, 19

4. 23, 26, 50

5. 32, 41, 63

6. 2.7, 3.1, 4.3

7. 0.7, 1.4, 2.1

8. 12.3, 13.9, 25.2

Find the range for the measure of the third side of a triangle given the measures of two sides.

9. 6 ft and 19 ft

10. 7 km and 29 km

11. 13 in. and 27 in.

12. 18 ft and 23 ft

13. 25 yd and 38 yd

14. 31 cm and 39 cm

15. 42 m and 6 m

16. 54 in. and 7 in.

17. Given: H is the centroid of $\triangle EDF$.

Prove: $EY + FY > DE$

Proof:

Statements	Reasons
1. H is the centroid of $\triangle EDF$.	1. Given
2. $\overline{EY}$ is a median.	2. _____
3. _____	3. Definition of median
4. _____	4. Definition of midpoint
5. $EY + DY > DE$	5. _____
6. $EY + FY > DE$	6. _____

18. GARDENING Ha Poong has 4 lengths of wood from which he plans to make a border for a triangular-shaped herb garden. The lengths of the wood borders are 8 inches, 10 inches, 12 inches, and 18 inches. How many different triangular borders can Ha Poong make?

5-6 Skills Practice

Inequalities Involving Two Triangles

Compare the given measures.

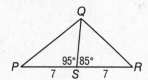

1. $m\angle BXA$ and $m\angle DXA$

2. BC and DC

Compare the given measures.

3. $m\angle STR$ and $m\angle TRU$

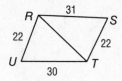

4. PQ and RQ

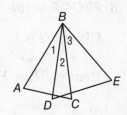

5. In the figure, $\overline{BA}$, $\overline{BD}$, $\overline{BC}$, and $\overline{BE}$ are congruent and $AC < DE$. How does $m\angle 1$ compare with $m\angle 3$? Explain your thinking.

6. PROOF Write a two-column proof.

Given: $\overline{BA} \cong \overline{DA}$
    $BC > DC$

Prove: $m\angle 1 > m\angle 2$

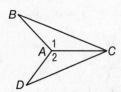

Proof:

Statements	Reasons
1. $BA \cong DA$	1. Given
2. $BC > DC$	2. Given
3. $AC \cong AC$	3. Reflexive Property
4. $m\angle 1 > m\angle 3$	4. SSS Inequality

5-6 Practice

Inequalities in Two Triangles

Compare the given measures.

1. *AB* and *BK*

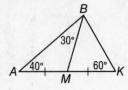

2. *ST* and *SR*

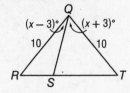

3. $m\angle CDF$ and $m\angle EDF$

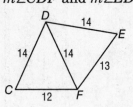

4. $m\angle R$ and $m\angle T$

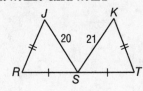

5. **PROOF** Write a two-column proof.

   Given: *G* is the midpoint of $\overline{DF}$.
   $m\angle 1 > m\angle 2$

   Prove: *ED* > *EF*

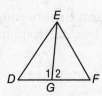

6. **TOOLS** Rebecca used a spring clamp to hold together a chair leg she repaired with wood glue. When she opened the clamp, she noticed that the angle between the handles of the clamp decreased as the distance between the handles of the clamp decreased. At the same time, the distance between the gripping ends of the clamp increased. When she released the handles, the distance between the gripping end of the clamp decreased and the distance between the handles increased. Is the clamp an example of the Hinge Theorem or its converse?

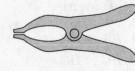

6-1 Skills Practice

Angles of Polygons

Find the sum of the measures of the interior angles of each convex polygon.

1. nonagon

2. heptagon

3. decagon

The measure of an interior angle of a regular polygon is given. Find the number of sides in the polygon.

4. 108

5. 120

6. 150

Find the measure of each interior angle.

7.

Parallelogram $ABCD$ with angles: $A = (2x - 15)°$, $B = x°$, $D = x°$, $C = (2x - 15)°$

8.

Quadrilateral $LMNP$ with angles: $L = (2x + 20)°$, $M = (3x - 10)°$, $P = (2x)°$, $N = (2x - 10)°$

9.

Trapezoid $STUW$ with angles: $S = (2x + 16)°$, $T = (2x + 16)°$, $W = (x + 14)°$, $U = (x + 14)°$

10.

Hexagon $DEFGHI$ with angles: $D = (7x)°$, $E = (7x)°$, $I = (4x)°$, $F = (4x)°$, $H = (7x)°$, $G = (7x)°$

Find the measures of each interior angle of each regular polygon.

11. quadrilateral

12. pentagon

13. dodecagon

Find the measures of each exterior angle of each regular polygon.

14. octagon

15. nonagon

16. 12-gon

6-1　Practice

Angles of Polygons

Find the sum of the measures of the interior angles of each convex polygon.

1. 11-gon　　　　　　　**2.** 14-gon　　　　　　　**3.** 17-gon

The measure of an interior angle of a regular polygon is given. Find the number of sides in the polygon.

4. 144　　　　　　　　**5.** 156　　　　　　　　**6.** 160

Find the measure of each interior angle.

7.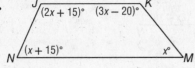

8.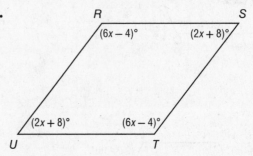

Find the measures of an exterior angle and an interior angle given the number of sides of each regular polygon. Round to the nearest tenth, if necessary.

9. 16　　　　　　　　**10.** 24　　　　　　　　**11.** 30

12. 14　　　　　　　　**13.** 22　　　　　　　　**14.** 40

15. CRYSTALLOGRAPHY Crystals are classified according to seven crystal systems. The basis of the classification is the shapes of the faces of the crystal. Turquoise belongs to the triclinic system. Each of the six faces of turquoise is in the shape of a parallelogram. Find the sum of the measures of the interior angles of one such face.

6-2 Skills Practice

Parallelograms

ALGEBRA Find the value of each variable.

1.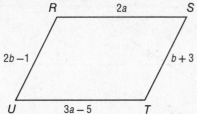

2.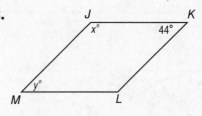

3.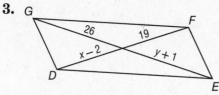

4.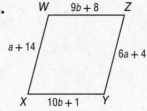

5.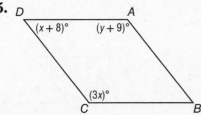

6.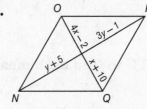

COORDINATE GEOMETRY Find the coordinates of the intersection of the diagonals of □HJKL with the given vertices.

7. $H(1, 1)$, $J(2, 3)$, $K(6, 3)$, $L(5, 1)$

8. $H(-1, 4)$, $J(3, 3)$, $K(3, -2)$, $L(-1, -1)$

9. **PROOF** Write a paragraph proof of the theorem *Consecutive angles in a parallelogram are supplementary*.

6-2 Practice

Parallelograms

ALGEBRA Find the value of each variable.

1.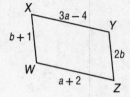

2.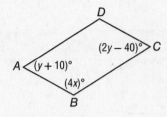

3.

4.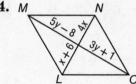

ALGEBRA Use $\square RSTU$ to find each measure or value.

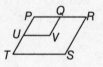

5. $m\angle RST =$ _____

6. $m\angle STU =$ _____

7. $m\angle TUR =$ _____

8. $b =$ _____

COORDINATE GEOMETRY Find the coordinates of the intersection of the diagonals of $\square PRYZ$ with the given vertices.

9. $P(2, 5)$, $R(3, 3)$, $Y(-2, -3)$, $Z(-3, -1)$

10. $P(2, 3)$, $R(1, -2)$, $Y(-5, -7)$, $Z(-4, -2)$

11. **PROOF** Write a paragraph proof of the following.
    Given: $\square PRST$ and $\square PQVU$
    Prove: $\angle V \cong \angle S$

12. **CONSTRUCTION** Mr. Rodriquez used the parallelogram at the right to design a herringbone pattern for a paving stone. He will use the paving stone for a sidewalk. If $m\angle 1$ is 130, find $m\angle 2$, $m\angle 3$, and $m\angle 4$.

6-3 Skills Practice

Tests for Parallelograms

Determine whether each quadrilateral is a parallelogram. Justify your answer.

1.

2.

3.

4.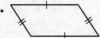

COORDINATE GEOMETRY Graph each quadrilateral with the given vertices. Determine whether the figure is a parallelogram. Justify your answer with the method indicated.

5. $P(0, 0)$, $Q(3, 4)$, $S(7, 4)$, $Y(4, 0)$; Slope Formula

6. $S(-2, 1)$, $R(1, 3)$, $T(2, 0)$, $Z(-1, -2)$; Distance and Slope Formulas

7. $W(2, 5)$, $R(3, 3)$, $Y(-2, -3)$, $N(-3, 1)$; Midpoint Formula

ALGEBRA Find x and y so that each quadrilateral is a parallelogram.

8.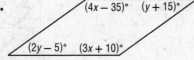

9.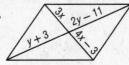

10.

11.

6-3 Practice

Tests for Parallelograms

Determine whether each quadrilateral is a parallelogram. Justify your answer.

1.

2.

3.

4.

COORDINATE GEOMETRY Graph each quadrilateral with the given vertices. Determine whether the figure is a parallelogram. Justify your answer with the method indicated.

5. $P(-5, 1)$, $S(-2, 2)$, $F(-1, -3)$, $T(2, -2)$; Slope Formula

6. $R(-2, 5)$, $O(1, 3)$, $M(-3, -4)$, $Y(-6, -2)$; Distance and Slope Formulas

ALGEBRA Find x and y so that the quadrilateral is a parallelogram.

7.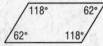
$(5x + 29)°$ $(5y - 9)°$
$(3y + 15)°$ $(7x - 11)°$

8.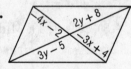
$-4x - 2$ $2y + 8$
$3y - 5$ $-3x + 4$

9.
$-6x$
$7y + 3$ $12y - 7$
$-4x + 6$

10.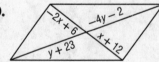
$-2x + 6$ $-4y - 2$
$y + 23$ $x + 12$

11. **TILE DESIGN** The pattern shown in the figure is to consist of congruent parallelograms. How can the designer be certain that the shapes are parallelograms?

6-4 Skills Practice

Rectangles

ALGEBRA Quadrilateral $ABCD$ is a rectangle.

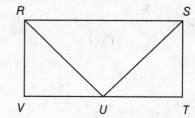

1. If $AC = 2x + 13$ and $DB = 4x - 1$, find DB.

2. If $AC = x + 3$ and $DB = 3x - 19$, find AC.

3. If $AE = 3x + 3$ and $EC = 5x - 15$, find AC.

4. If $DE = 6x - 7$ and $AE = 4x + 9$, find DB.

5. If $m\angle DAC = 2x + 4$ and $m\angle BAC = 3x + 1$, find $m\angle BAC$.

6. If $m\angle BDC = 7x + 1$ and $m\angle ADB = 9x - 7$, find $m\angle BDC$.

7. If $m\angle ABD = 7x - 31$ and $m\angle CDB = 4x + 5$, find $m\angle ABD$.

8. If $m\angle BAC = x + 3$ and $m\angle CAD = x + 15$, find $m\angle BAC$.

9. PROOF: Write a two-column proof.

Given: $RSTV$ is a rectangle and U is the midpoint of $\overline{VT}$.

Prove: $\triangle RUV \cong \triangle SUT$

Statements	Reasons

COORDINATE GEOMETRY Graph each quadrilateral with the given vertices. Determine whether the figure is a rectangle. Justify your answer using the indicated formula.

10. $P(-3, -2)$, $Q(-4, 2)$, $R(2, 4)$, $S(3, 0)$; Slope Formula

11. $J(-6, 3)$, $K(0, 6)$, $L(2, 2)$, $M(-4, -1)$; Distance Formula

12. $T(4, 1)$, $U(3, -1)$, $X(-3, 2)$, $Y(-2, 4)$; Distance Formula

6-4 Practice

Rectangles

ALGEBRA Quadrilateral *RSTU* is a rectangle.

1. If $UZ = x + 21$ and $ZS = 3x - 15$, find *US*.

2. If $RZ = 3x + 8$ and $ZS = 6x - 28$, find *UZ*.

3. If $RT = 5x + 8$ and $RZ = 4x + 1$, find *ZT*.

4. If $m\angle SUT = 3x + 6$ and $m\angle RUS = 5x - 4$, find $m\angle SUT$.

5. If $m\angle SRT = x + 9$ and $m\angle UTR = 2x - 44$, find $m\angle UTR$.

6. If $m\angle RSU = x + 41$ and $m\angle TUS = 3x + 9$, find $m\angle RSU$.

Quadrilateral *GHJK* is a rectangle. Find each measure if $m\angle 1 = 37$.

7. $m\angle 2$ _____

8. $m\angle 3$

9. $m\angle 4$

10. $m\angle 5$

11. $m\angle 6$

12. $m\angle 7$

COORDINATE GEOMETRY Graph each quadrilateral with the given vertices. Determine whether the figure is a rectangle. Justify your answer using the indicated formula.

13. $B(-4, 3)$, $G(-2, 4)$, $H(1, -2)$, $L(-1, -3)$; Slope Formula

14. $N(-4, 5)$, $O(6, 0)$, $P(3, -6)$, $Q(-7, -1)$; Distance Formula

15. $C(0, 5)$, $D(4, 7)$, $E(5, 4)$, $F(1, 2)$; Slope Formula

16. **LANDSCAPING** Huntington Park officials approved a rectangular plot of land for a Japanese Zen garden. Is it sufficient to know that opposite sides of the garden plot are congruent and parallel to determine that the garden plot is rectangular? Explain.

6-5 Skills Practice

Rhombi and Squares

ALGEBRA Quadrilateral *DKLM* is a rhombus.

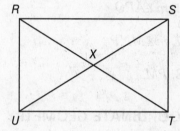

1. If $DK = 8$, find KL.

2. If $m\angle DML = 82$ find $m\angle DKM$.

3. If $m\angle KAL = 2x - 8$, find x.

4. If $DA = 4x$ and $AL = 5x - 3$, find DL.

5. If $DA = 4x$ and $AL = 5x - 3$, find AD.

6. If $DM = 5y + 2$ and $DK = 3y + 6$, find KL.

7. **PROOF** Write a two-column proof.
   Given: *RSTU* is a parallelogram.
   $\overline{RX} \cong \overline{TX} \cong \overline{SX} \cong \overline{UX}$
   Prove: *RSTU* is a rectangle.

Statements	Reasons

COORDINATE GEOMETRY Given each set of vertices, determine whether $\square QRST$ is a *rhombus*, a *rectangle*, or a *square*. List all that apply. Explain.

8. $Q(3, 5)$, $R(3, 1)$, $S(-1, 1)$, $T(-1, 5)$

9. $Q(-5, 12)$, $R(5, 12)$, $S(-1, 4)$, $T(-11, 4)$

10. $Q(-6, -1)$, $R(4, -6)$, $S(2, 5)$, $T(-8, 10)$

11. $Q(2, -4)$, $R(-6, -8)$, $S(-10, 2)$, $T(-2, 6)$

6-5 Practice

Rhombi and Squares

PRYZ is a rhombus. If *RK* = 5, *RY* = 13 and *m∠YRZ* = 67, find each measure.

1. *KY*

2. *PK*

3. *m∠YKZ*

4. *m∠PZR*

MNPQ is a rhombus. If *PQ* = $3\sqrt{2}$ and *AP* = 3, find each measure.

5. *AQ*

6. *m∠APQ*

7. *m∠MNP*

8. *PM*

COORDINATE GEOMETRY Given each set of vertices, determine whether ☐*BEFG* is a *rhombus*, a *rectangle*, or a *square*. List all that apply. Explain.

9. *B*(−9, 1), *E*(2, 3), *F*(12, −2), *G*(1, −4)

10. *B*(1, 3), *E*(7, −3), *F*(1, −9), *G*(−5, −3)

11. *B*(−4, −5), *E*(1, −5), *F*(−2, −1), *G*(−7, −1)

12. **TESSELLATIONS** The figure is an example of a tessellation. Use a ruler or protractor to measure the shapes and then name the quadrilaterals used to form the figure.

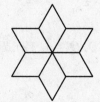

6-6   Skills Practice

Trapezoids and Kites

ALGEBRA Find each measure.

1. $m\angle S$

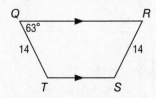

2. $m\angle M$

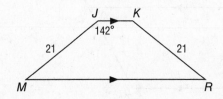

3. $m\angle D$

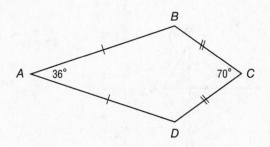

4. RH

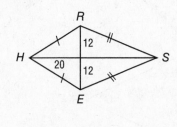

ALGEBRA For trapezoid $HJKL$, T and S are midpoints of the legs.

5. If $HJ = 14$ and $LK = 42$, find TS.

6. If $LK = 19$ and $TS = 15$, find HJ.

7. If $HJ = 7$ and $TS = 10$, find LK.

8. If $KL = 17$ and $JH = 9$, find ST.

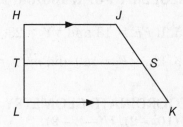

COORDINATE GEOMETRY $EFGH$ is a quadrilateral with vertices $E(1, 3)$, $F(5, 0)$, $G(8, -5)$, $H(-4, 4)$.

9. Verify that $EFGH$ is a trapezoid.

10. Determine whether $EFGH$ is an isosceles trapezoid. Explain.

6-6 Practice

Trapezoids and Kites

Find each measure.

1. $m\angle T$

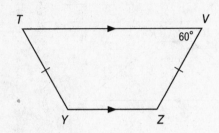

2. $m\angle Y$

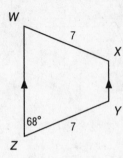

3. $m\angle Q$

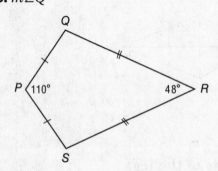

4. BC

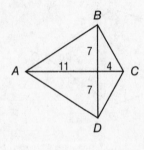

ALGEBRA For trapezoid *FEDC*, *V* and *Y* are midpoints of the legs.

5. If $FE = 18$ and $VY = 28$, find CD.

6. If $m\angle F = 140$ and $m\angle E = 125$, find $m\angle D$.

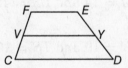

COORDINATE GEOMETRY *RSTU* is a quadrilateral with vertices $R(-3, -3)$, $S(5, 1)$, $T(10, -2)$, $U(-4, -9)$.

7. Verify that *RSTU* is a trapezoid.

8. Determine whether *RSTU* is an isosceles trapezoid. Explain.

9. CONSTRUCTION A set of stairs leading to the entrance of a building is designed in the shape of an isosceles trapezoid with the longer base at the bottom of the stairs and the shorter base at the top. If the bottom of the stairs is 21 feet wide and the top is 14 feet wide, find the width of the stairs halfway to the top.

10. DESK TOPS A carpenter needs to replace several trapezoid-shaped desktops in a classroom. The carpenter knows the lengths of both bases of the desktop. What other measurements, if any, does the carpenter need?

7-1 Skills Practice

Ratios and Proportions

1. **FOOTBALL** A tight end scored 6 touchdowns in 14 games. Find the ratio of touchdowns per game.

2. **EDUCATION** In a schedule of 6 classes, Marta has 2 elective classes. What is the ratio of elective to non-elective classes in Marta's schedule?

3. **BIOLOGY** Out of 274 listed species of birds in the United States, 78 species made the endangered list. Find the ratio of endangered species of birds to listed species in the United States.

4. **BOARD GAMES** Myra is playing a board game. After 12 turns, Myra has landed on a blue space 3 times. If the game will last for 100 turns, predict how many times Myra will land on a blue space.

5. **SCHOOL** The ratio of male students to female students in the drama club at Campbell High School is 3:4. If the number of male students in the club is 18, predict the number of female students?

Solve each proportion.

6. $\dfrac{2}{5} = \dfrac{x}{40}$

7. $\dfrac{7}{10} = \dfrac{21}{x}$

8. $\dfrac{20}{5} = \dfrac{4x}{6}$

9. $\dfrac{5x}{4} = \dfrac{35}{8}$

10. $\dfrac{x+1}{3} = \dfrac{7}{2}$

11. $\dfrac{15}{3} = \dfrac{x-3}{5}$

12. The ratio of the measures of the sides of a triangle is 3:5:7, and its perimeter is 450 centimeters. Find the measures of each side of the triangle.

13. The ratio of the measures of the sides of a triangle is 5:6:9, and its perimeter is 220 meters. What are the measures of the sides of the triangle?

14. The ratio of the measures of the sides of a triangle is 4:6:8, and its perimeter is 126 feet. What are the measures of the sides of the triangle?

15. The ratio of the measures of the sides of a triangle is 5:7:8, and its perimeter is 40 inches. Find the measures of each side of the triangle.

7-1 Practice

Ratios and Proportions

1. **NUTRITION** One ounce of cheddar cheese contains 9 grams of fat. Six of the grams of fat are saturated fats. Find the ratio of saturated fats to total fat in an ounce of cheese.

2. **FARMING** The ratio of goats to sheep at a university research farm is 4:7. The number of sheep at the farm is 28. What is the number of goats?

3. **QUALITY CONTROL** A worker at an automobile assembly plant checks new cars for defects. Of the first 280 cars he checks, 4 have defects. If 10,500 cars will be checked this month, predict the total number of cars that will have defects.

Solve each proportion.

4. $\dfrac{5}{8} = \dfrac{x}{12}$

5. $\dfrac{x}{1.12} = \dfrac{1}{5}$

6. $\dfrac{6x}{27} = 43$

7. $\dfrac{x+2}{3} = \dfrac{8}{9}$

8. $\dfrac{3x-5}{4} = \dfrac{-5}{7}$

9. $\dfrac{x-2}{4} = \dfrac{x+4}{2}$

10. The ratio of the measures of the sides of a triangle is 3:4:6, and its perimeter is 104 feet. Find the measure of each side of the triangle.

11. The ratio of the measures of the sides of a triangle is 7:9:12, and its perimeter is 84 inches. Find the measure of each side of the triangle.

12. The ratio of the measures of the sides of a triangle is 6:7:9, and its perimeter is 77 centimeters. Find the measure of each side of the triangle.

13. The ratio of the measures of the three angles is 4:5:6. Find the measure of each angle of the triangle.

14. The ratio of the measures of the three angles is 5:7:8. Find the measure of each angle of the triangle.

15. **BRIDGES** A construction worker is placing rivets in a new bridge. He uses 42 rivets to build the first 2 feet of the bridge. If the bridge is to be 2200 feet in length, predict the number of rivets that will be needed for the entire bridge.

7-2 Skills Practice

Similar Polygons

Determine whether each pair of figures is similar. If so, write the similarity statement and scale factor. If not, explain your reasoning.

1.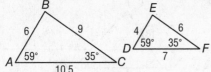

2.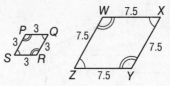

Each pair of polygons is similar. Find the value of *x*.

3.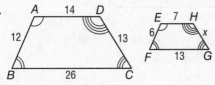

4.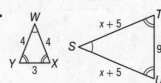

5.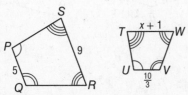

6.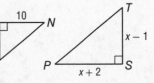

7-2   Practice

Similar Polygons

Determine whether each pair of figures is similar. If so, write the similarity statement and scale factor. If not, explain your reasoning.

1.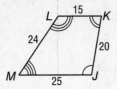

2.

Each pair of polygons is similar. Find the value of x.

3.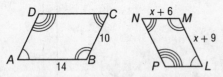

4.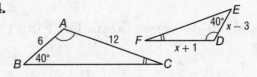

5. **PENTAGONS** If $ABCDE \sim PQRST$, find the scale factor of $ABCDE$ to $PQRST$ and the perimeter of each polygon.

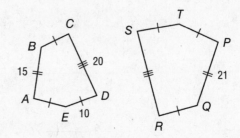

6. **SWIMMING POOLS** The Minnitte family and the neighboring Gaudet family both have in-ground swimming pools. The Minnitte family pool, $PQRS$, measures 48 feet by 84 feet. The Gaudet family pool, $WXYZ$, measures 40 feet by 70 feet. Are the two pools similar? If so, write the similarity statement and scale factor.

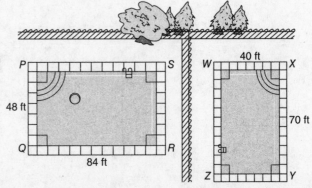

7-3 Skills Practice

Similar Triangles

Determine whether each pair of triangles is similar. If so, write a similarity statement. If not, what would be sufficient to prove the triangles similar? Explain your reasoning.

1.

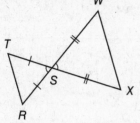

2.

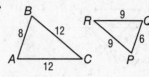

3.

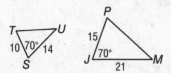

4.

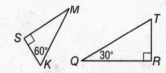

ALGEBRA Identify the similar triangles. Then find each measure.

5. *AC*

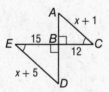

6. *JL*

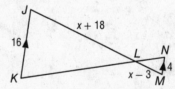

7. *EH*

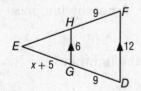

8. *VT*

7-3 Practice

Similar Triangles

Determine whether the triangles are similar. If so, write a similarity statement. If not, what would be sufficient to prove the triangles similar? Explain your reasoning.

1.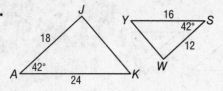

2.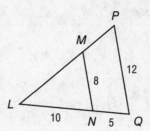

ALGEBRA Identify the similar triangles. Then find each measure.

3. *LM, QP*

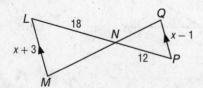

4. *NL, ML*

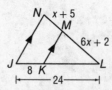

5. *PS, PR*

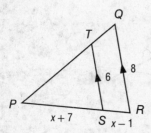

6. *EG, HG*

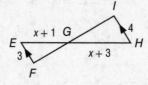

7. **INDIRECT MEASUREMENT** A lighthouse casts a 128-foot shadow. A nearby lamppost that measures 5 feet 3 inches casts an 8-foot shadow.

   a. Write a proportion that can be used to determine the height of the lighthouse.

   b. What is the height of the lighthouse?

7-4 Skills Practice

Parallel Lines and Proportional Parts

1. If $JK = 7$, $KH = 21$, and $JL = 6$, find LI.

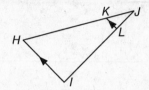

2. If $RU = 8$, $US = 14$, $TV = x - 1$, and $VS = 17.5$, find x and TV.

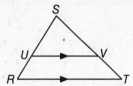

Determine whether $\overline{BC} \parallel \overline{DE}$. Justify your answer.

3. $AD = 15$, $DB = 12$, $AE = 10$, and $EC = 8$

4. $BD = 9$, $BA = 27$, and $CE = \frac{1}{3}EA$

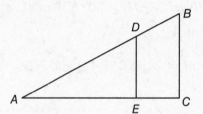

5. $AE = 30$, $AC = 45$, and $AD = 2DB$

$\overline{JH}$ **is a midsegment of $\triangle KLM$. Find the value of x.**

6.

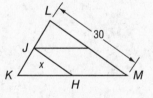

7.

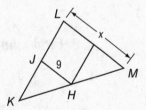

8.

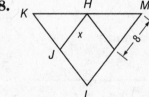

ALGEBRA Find x and y.

9.

10.

7-4 Practice

Parallel Lines and Proportional Parts

1. If $AD = 24$, $DB = 27$, and $EB = 18$, find CE.

2. If $QT = x + 6$, $SR = 12$, $PS = 27$, and $TR = x - 4$, find QT and TR.

Determine whether $\overline{JK} \parallel \overline{NM}$. Justify your answer.

3. $JN = 18$, $JL = 30$, $KM = 21$, and $ML = 35$

4. $KM = 24$, $KL = 44$, and $NL = \frac{5}{6} JN$

$\overline{JH}$ is a midsegment of $\triangle KLM$. Find the value of x.

5.

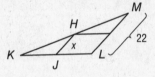

6.

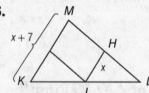

7. Find x and y.

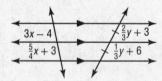

8. Find x and y.

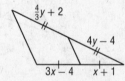

9. MAPS On a map, Wilmington Street, Beech Drive, and Ash Grove Lane appear to all be parallel. The distance from Wilmington to Ash Grove along Kendall is 820 feet and along Magnolia, 660 feet. If the distance between Beech and Ash Grove along Magnolia is 280 feet, what is the distance between the two streets along Kendall?

7-5 Skills Practice

Parts of Similar Triangles

Find x.

1.
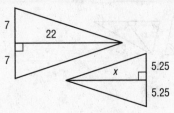

2.
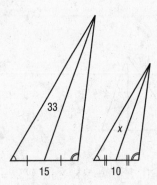

3.
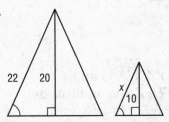

4.
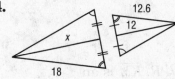

5. If $\triangle RST \sim \triangle EFG$, $\overline{SH}$ is an altitude of $\triangle RST$, $\overline{FJ}$ is an altitude of $\triangle EFG$, $ST = 6$, $SH = 5$, and $FJ = 7$, find FG.

6. If $\triangle ABC \sim \triangle MNP$, $\overline{AD}$ is an altitude of $\triangle ABC$, $\overline{MQ}$ is an altitude of $\triangle MNP$, $AB = 24$, $AD = 14$, and $MQ = 10.5$, find MN.

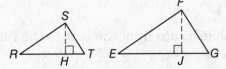

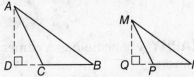

Find the value of each variable.

7.
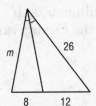

8.
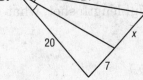

7-5 Practice

Parts of Similar Triangles

ALGEBRA Find x.

1.

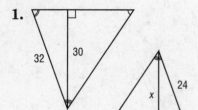

2.

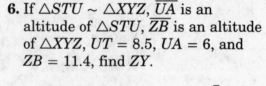

3.

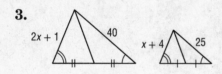

4.

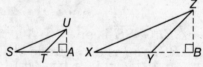

5. If $\triangle JKL \sim \triangle NPR$, $\overline{KM}$ is an altitude of $\triangle JKL$, $\overline{PT}$ is an altitude of $\triangle NPR$, $KL = 28$, $KM = 18$, and $PT = 15.75$, find PR.

6. If $\triangle STU \sim \triangle XYZ$, $\overline{UA}$ is an altitude of $\triangle STU$, $\overline{ZB}$ is an altitude of $\triangle XYZ$, $UT = 8.5$, $UA = 6$, and $ZB = 11.4$, find ZY.

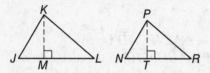

7. PHOTOGRAPHY Francine has a camera in which the distance from the lens to the film is 24 millimeters.

a. If Francine takes a full-length photograph of her friend from a distance of 3 meters and the height of her friend is 140 centimeters, what will be the height of the image on the film? (*Hint*: Convert to the same unit of measure.)

b. Suppose the height of the image on the film of her friend is 15 millimeters. If Francine took a full-length shot, what was the distance between the camera and her friend?

7-6 Skills Practice

Similarity Transformations

Determine whether the dilation from *A* to *B* is an *enlargement* or a *reduction*. Then find the scale factor of the dilation.

1.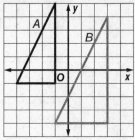

2.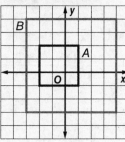

3.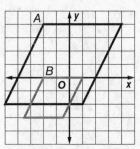

4.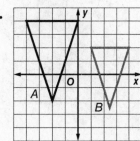

Graph the original figure and its dilated image. Then verify that the dilation is a similarity transformation.

5. $A(-3, 4)$, $B(3, 4)$, $C(-3, -2)$; $A'(-2, 3)$, $B'(0, 3)$, $C'(-2, 1)$

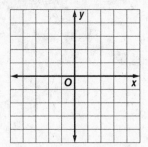

6. $F(-3, 4)$, $G(2, 4)$, $H(2, -2)$, $J(-3, -2)$; $F'(-1.5, 3)$, $G'(1, 3)$, $H'(1, 0)$, $J'(-1.5, 0)$

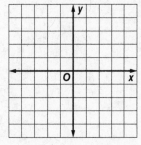

7. $P(-3, 1)$, $Q(-1, 1)$, $R(-1, -3)$; $P'(-1, 4)$, $Q'(3, 4)$, $R'(3, -4)$

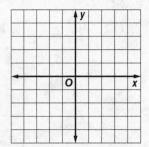

8. $A(-5, -1)$, $B(0, 1)$, $C(5, -1)$ $A'(1, -1.5)$, $B'(2, 0)$, $C'(3, -1.5)$

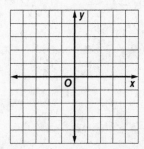

7-6 **Practice**

Similarity Transformations

Determine whether the dilation from *A* to *B* is an *enlargement* or a *reduction*.
Then find the scale factor of the dilation.

1.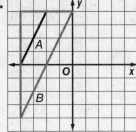

2.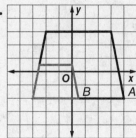

3.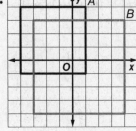

4.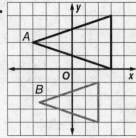

5.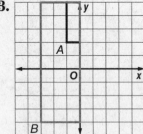

6.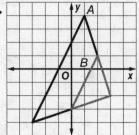

Graph the original figure and its dilated image. Then verify that the dilation is a similarity transformation.

7. *Q*(1, 4), *R*(4, 4), *S*(4, −1),

   X(−4, 5), *Y*(2, 5), *Z*(2, −5)

8. *A*(−4, 2), *B*(0, 4), *C*(4, 2);

   F(−2, 1), *G*(0, 2), *H*(2, 1)

7-7 Skills Practice

Scale Drawings and Models

MAPS Use the map shown and a customary ruler to find the actual distance between each pair of cities. Measure to the nearest sixteenth of an inch.

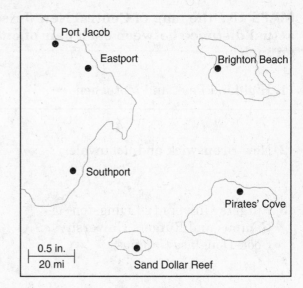

1. Port Jacob and Southport

2. Port Jacob and Brighton Beach

3. Brighton Beach and Pirates' Cove

4. Eastport and Sand Dollar Reef

5. **SCALE MODEL** Sanjay is making a 139 centimeters long scale model of the Parthenon for his World History class. The actual length of the Parthenon is 69.5 meters long.

   a. What is the scale of the model?

   b. How many times as long as the actual Parthenon is the model?

6. **ARCHITECTURE** An architect is making a scale model of an office building he wishes to construct. The model is 9 inches tall. The actual office building he plans to construct will be 75 feet tall.

   a. What is the scale of the model?

   b. What scale factor did the architect use to build his model?

7. **WHITE HOUSE** Craig is making a scale drawing of the White House on an 8.5-by-11-inch sheet of paper. The White House is 168 feet long and 152 feet wide. Choose an appropriate scale for the drawing and use that scale to determine the drawing's dimensions.

8. **GEOGRAPHY** Choose an appropriate scale and construct a scale drawing of each rectangular state to fit on a 4-by-6-inch index card.
   a. The state of Colorado is approximately 380 miles long (east to west) and 280 miles wide (north to south).

   b. The state of Wyoming is approximately 365 miles long (east to west) and 265 miles wide (north to south).

7-7 Practice

Scale Drawings and Models

MAPS Use the map of Central New Jersey shown and an inch ruler to find the actual distance between each pair of cities. Measure to the nearest sixteenth of an inch.

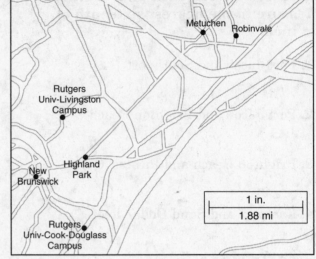

1. Highland Park and Metuchen

2. New Brunswick and Robinvale

3. Rutgers University Livingston Campus and Rutgers University Cook-Douglass Campus

4. AIRPLANES William is building a scale model of a Boeing 747-400 aircraft. The wingspan of the model is approximately 8 feet $10\frac{1}{16}$ inches. If the scale factor of the model is approximately 1:24, what is the actual wingspan of a Boeing 747-400 aircraft?

5. ENGINEERING A civil engineer is making a scale model of a highway on ramp. The length of the model is 4 inches. The actual length of the on ramp is 500 feet.

  a. What is the scale of the model?

  b. How many times as long as the actual on ramp is the model?

  c. How many times as long as the model is the actual on ramp?

6. MOVIES A movie director is creating a scale model of the Empire State Building to use in a scene. The Empire State Building is 1250 feet tall.

  a. If the model is 75 inches tall, what is the scale of the model?

  b. How tall would the model be if the director uses a scale factor of 1:75?

7. MONA LISA A visitor to the Louvre Museum in Paris wants to sketch a drawing of the *Mona Lisa,* a famous painting. The original painting is 77 centimeters by 53 centimeters. Choose an appropriate scale for the replica so that it will fit on a 8.5-by-11-inch sheet of paper.

8-1 Skills Practice

Geometric Mean

Find the geometric mean between each pair of numbers.

1. 2 and 8

2. 9 and 36

3. 4 and 7

4. 5 and 10

5. 28 and 14

6. 7 and 36

Write a similarity statement identifying the three similar triangles in the figure.

7.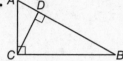

8.

9.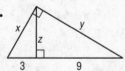

10.

Find *x, y* and *z*.

11.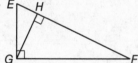

12.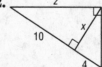

13.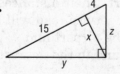

14.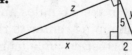

8-1 Practice

Geometric Mean

Find the geometric mean between each pair of numbers.

1. 8 and 12

2. 3 and 15

3. $\frac{4}{5}$ and 2

Write a similarity statement identifying the three similar triangles in the figure.

4.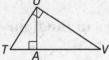

5.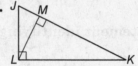

Find x, y, and z.

6.

7.

8.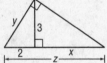

9.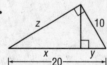

10. CIVIL An airport, a factory, and a shopping center are at the vertices of a right triangle formed by three highways. The airport and factory are 6.0 miles apart. Their distances from the shopping center are 3.6 miles and 4.8 miles, respectively. A service road will be constructed from the shopping center to the highway that connects the airport and factory. What is the shortest possible length for the service road? Round to the nearest hundredth.

8-2 Skills Practice

The Pythagorean Theorem and Its Converse

Find x.

1.

2.
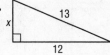

3.
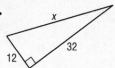

4.
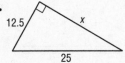

5.

6.
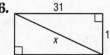

Use a Pythagorean Triple to find x.

7.
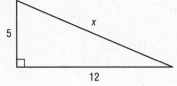

8.
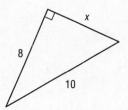

9.
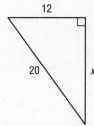

10.
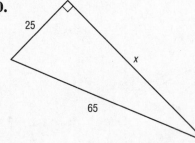

11.
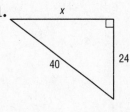

12.
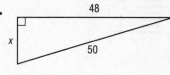

Determine whether each set of numbers can be measure of the sides of a triangle. If so, classify the triangle as *acute, obtuse, or right*. Justify your answer.

13. 7, 24, 25   **14.** 8, 14, 20   **15.** 12.5, 13, 26

16. $3\sqrt{2}$, $\sqrt{7}$, 4   **17.** 20, 21, 29   **18.** 32, 35, 70

8-2 Practice

The Pythagorean Theorem and Its Converse

Find x.

1.

2.

3.

4.

5.
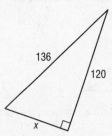

6.
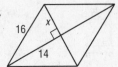

Use a Pythagorean Triple to find x.

7.

8.
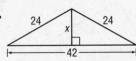

9.
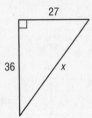

10.
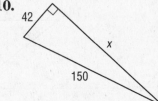

Determine whether each set of numbers can be measure of the sides of a triangle. If so, classify the triangle as *acute, obtuse, or right*. Justify your answer.

11. 10, 11, 20

12. 12, 14, 49

13. $5\sqrt{2}$, 10, 11

14. 21.5, 24, 55.5

15. 30, 40, 50

16. 65, 72, 97

17. CONSTRUCTION The bottom end of a ramp at a warehouse is 10 feet from the base of the main dock and is 11 feet long. How high is the dock?

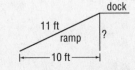

8-3 Skills Practice

Special Right Triangles

Find x.

1.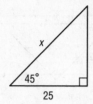

2.

3.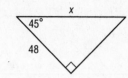

4.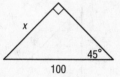

5.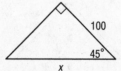

6.

7. Determine the length of the leg of 45°-45°-90° triangle with a hypotenuse length of 26.

8. Find the length of the hypotenuse of a 45°-45°-90° triangle with a leg length of 50 centimeters.

Find x and y.

9.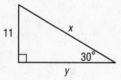

10.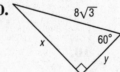

11.

12.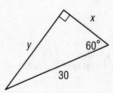

13.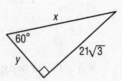

14.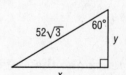

15. An equilateral triangle has an altitude length of 27 feet. Determine the length of a side of the triangle.

16. Find the length of the side of an equilateral triangle that has an altitude length of $11\sqrt{3}$ feet.

8-3 Practice

Special Right Triangles

Find x.

1.
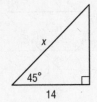

2.
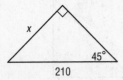

3.
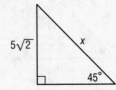

4.
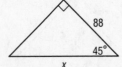

5.
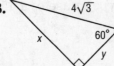

6.
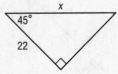

Find x and y.

7.
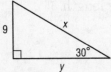

8.
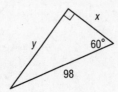

9.
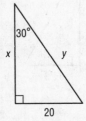

10.
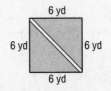

11. Determine the length of the leg of 45°-45°-90° triangle with a hypotenuse length of 38.

12. Find the length of the hypotenuse of a 45°-45°-90° triangle with a leg length of 77 centimeters.

13. An equilateral triangle has an altitude length of 33 feet. Determine the length of a side of the triangle.

14. BOTANICAL GARDENS One of the displays at a botanical garden is an herb garden planted in the shape of a square. The square measures 6 yards on each side. Visitors can view the herbs from a diagonal pathway through the garden. How long is the pathway?

8-4 Skills Practice

Trigonometry

Find sin R, cos R, tan R, sin S, cos S, and tan S.
Express each ratio as a fraction and as a decimal to the
nearest hundredth.

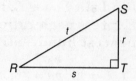

1. $r = 16, s = 30, t = 34$ **2.** $r = 10, s = 24, t = 26$

Use a special right triangle to express each trigonometric ratio as a fraction and as a decimal to the nearest hundredth if necessary.

3. sin 30°   **4.** tan 45°   **5.** cos 60°

6. sin 60°   **7.** tan 30°   **8.** cos 45°

Find x. Round to the nearest hundredth if necessary.

9.
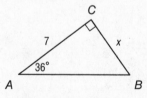

10.
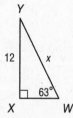

11.
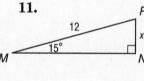

Use a calculator to find the measure of ∠B to the nearest tenth.

12.
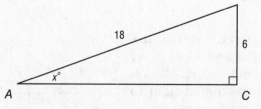

13.
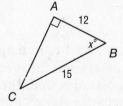

14.
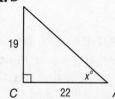

8-4 Practice

Trigonometry

Find sin L, cos L, tan L, sin M, cos M, and tan M.
Express each ratio as a fraction and as a decimal to the
nearest hundredth.

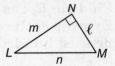

1. $\ell = 15$, $m = 36$, $n = 39$

2. $\ell = 12$, $m = 12\sqrt{3}$, $n = 24$

Find x. Round to the nearest hundredth.

3.

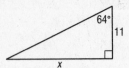

4.

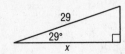

5.

Use a calculator to find the measure of $\angle B$ to the nearest tenth.

6.

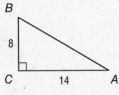

7.

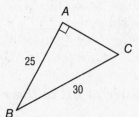

8.

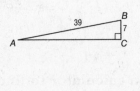

9. GEOGRAPHY Diego used a theodolite to map a region of land for his
class in geomorphology. To determine the elevation of a vertical rock
formation, he measured the distance from the base of the formation to
his position and the angle between the ground and the line of sight to the
top of the formation. The distance was 43 meters and the angle was
$36°$. What is the height of the formation to the nearest meter?

8-5 **Skills Practice**

Angles of Elevation and Depression

Name the angle of depression or angle of elevation in each figure.

1.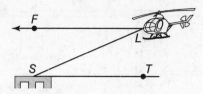

2.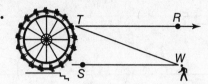

3.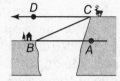

4.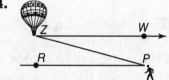

5. **MOUNTAIN BIKING** On a mountain bike trip along the Gemini Bridges Trail in Moab, Utah, Nabuko stopped on the canyon floor to get a good view of the twin sandstone bridges. Nabuko is standing about 60 meters from the base of the canyon cliff, and the natural arch bridges are about 100 meters up the canyon wall. If her line of sight is 5 metres above the ground, what is the angle of elevation to the top of the bridges? Round to the nearest tenth degree.

6. **SHADOWS** Suppose the sun casts a shadow off a 35-foot building. If the angle of elevation to the sun is 60°, how long is the shadow to the nearest tenth of a foot?

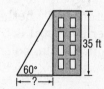

7. **BALLOONING** Angie sees a hot air balloon in the sky from her spot on the ground. The angle of elevation from Angie to the balloon is 40°. If she steps back 200 feet, the new angle of elevation is 10°. If Angie is 5.5 feet tall, how far off the ground is the hot air balloon?

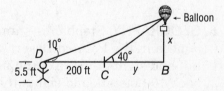

8. **INDIRECT MEASUREMENT** Kyle is at the end of a pier 30 feet above the ocean. His eye level is 3 feet above the pier. He is using binoculars to watch a whale surface. If the angle of depression of the whale is 20°, how far is the whale from Kyle's binoculars? Round to the nearest tenth foot.

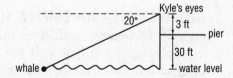

8-5 Practice

Angles of Elevation and Depression

Name the angle of depression or angle of elevation in each figure.

1.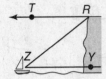

2.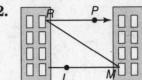

3. **WATER TOWERS** A student can see a water tower from the closest point of the soccer field at San Lobos High School. The edge of the soccer field is about 110 feet from the water tower and the water tower stands at a height of 32.5 feet. What is the angle of elevation if the eye level of the student viewing the tower from the edge of the soccer field is 6 feet above the ground? Round to the nearest tenth.

4. **CONSTRUCTION** A roofer props a ladder against a wall so that the top of the ladder reaches a 30-foot roof that needs repair. If the angle of elevation from the bottom of the ladder to the roof is 55°, how far is the ladder from the base of the wall? Round your answer to the nearest foot.

5. **TOWN ORDINANCES** The town of Belmont restricts the height of flagpoles to 25 feet on any property. Lindsay wants to determine whether her school is in compliance with the regulation. Her eye level is 5.5 feet from the ground and she stands 36 feet from the flagpole. If the angle of elevation is about 25°, what is the height of the flagpole to the nearest tenth?

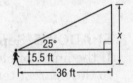

6. **GEOGRAPHY** Stephan is standing on the ground by a mesa in the Painted Desert. Stephan is 1.8 meters tall and sights the top of the mesa at 29°. Stephan steps back 100 meters and sights the top at 25°. How tall is the mesa?

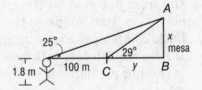

7. **INDIRECT MEASUREMENT** Mr. Dominguez is standing on a 40-foot ocean bluff near his home. He can see his two dogs on the beach below. If his line of sight is 6 feet above the ground and the angles of depression to his dogs are 34° and 48°, how far apart are the dogs to the nearest foot?

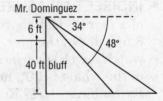

8-6   Skills Practice

The Law of Sines and Law of Cosines

Find *x*. Round angle measures to the nearest degree and side lengths to the nearest tenth.

1.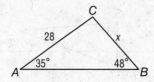

2.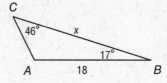

3.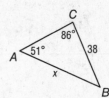

4.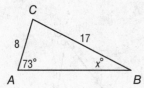

5.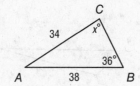

6.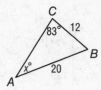

7.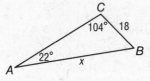

8.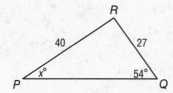

9.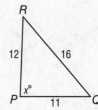

10.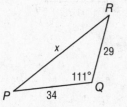

11.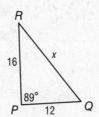

12.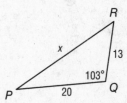

13.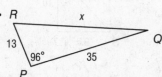

14.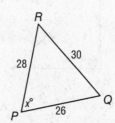

15.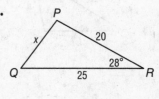

16. Solve the triangle. Round angle measures to the nearest degree.

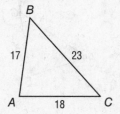

8-6 Practice

The Law of Sines and Law of Cosines

Find *x*. Round angle measures to the nearest degree and side lengths to the nearest tenth.

1.

2.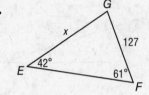

3.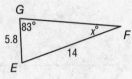

4.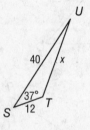

5.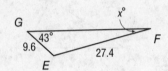

6.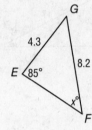

7.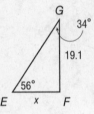

8.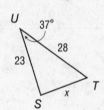

9.

10.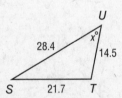

11.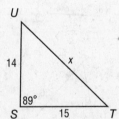

12.

13. **INDIRECT MEASUREMENT** To find the distance from the edge of the lake to the tree on the island in the lake, Hannah set up a triangular configuration as shown in the diagram. The distance from location *A* to location *B* is 85 meters. The measures of the angles at *A* and *B* are 51° and 83°, respectively. What is the distance from the edge of the lake at *B* to the tree on the island at *C*?

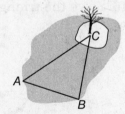

8-7 Skills Practice

Vectors

Write the component form of each vector.

1.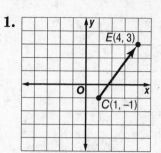

2.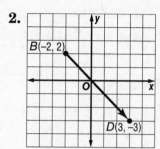

Find the magnitude and direction of each vector.

3. $\overrightarrow{LM}$: $L(2, -3)$, $M(4, 9)$

4. $\overrightarrow{PQ}$: $P(0, 2)$, $Q(3, 12)$

5. $\overrightarrow{KV}$: $K(5, 4)$, $V(-3, 1)$

6. $\overrightarrow{JK}$: $J(1, 5)$, $K(-4, -6)$

Copy the vectors to find each sum or difference.

7. $\vec{a} + \vec{z}$

8. $\vec{t} - \vec{r}$

9. $\vec{p} - \vec{q}$

10. $\vec{s} + \vec{t}$

109

8-7 Practice

Vectors

Write the component form of each vector.

1.

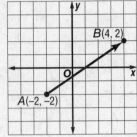

2.

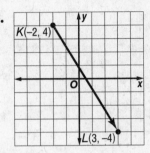

Find the magnitude and direction of each vector.

3. $\overrightarrow{ST}$: $S(-8, -5)$, $T(-2, 7)$

4. $\overrightarrow{FG}$: $F(-4, 1)$, $G(5, -6)$

Copy the vectors to find each sum or difference.

5. $\vec{p} + \vec{r}$

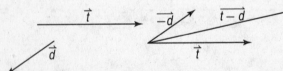

6. $\vec{a} - \vec{b}$

7. $\vec{t} - \vec{d}$

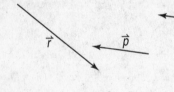

8. $\vec{c} + \vec{f}$

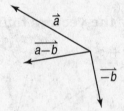

9. **AVIATION** A jet begins a flight along a path due north at 300 miles per hour. A wind is blowing due west at 30 miles per hour.

a. Find the resultant velocity of the plane.

b. Find the resultant direction of the plane.

9-1 Skills Practice

Reflections

Use the given figure and line of reflection. Draw the image in this line using a ruler.

1.

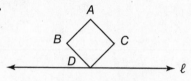

2.

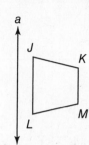

3.

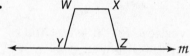

4.

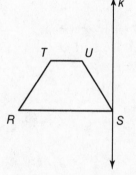

COORDINATE GEOMETRY Graph each figure and its image under the given reflection.

5. $\triangle ABC$ with vertices $A(-3, 2)$, $B(0, 1)$, and $C(-2, -3)$ in the line $y = x$

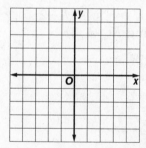

6. trapezoid $DEFG$ with vertices $D(0, -3)$, $E(1, 3)$, $F(3, 3)$, and $G(4, -3)$ in the y-axis

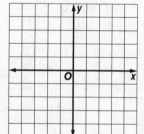

7. parallelogram $RSTU$ with vertices $R(-2, 3)$, $S(2, 4)$, $T(2, -3)$ and $U(-2, -4)$ in the line $y = x$

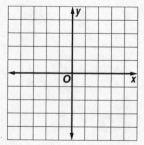

8. square $KLMN$ with vertices $K(-1, 0)$, $L(-2, 3)$, $M(1, 4)$, and $N(2, 1)$ in the x-axis

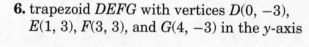

9-1 Practice

Reflections

Use the figure and given line of reflection. Then draw the reflected image in this line using a ruler.

1.

2.

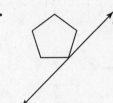

COORDINATE GEOMETRY Graph each figure and its image under the given reflection.

3. quadrilateral *ABCD* with vertices *A*(−3, 3), *B*(1, 4), *C*(4, 0), and *D*(−3, −3) in the line $y = x$

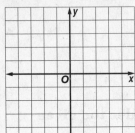

4. △*FGH* with vertices *F*(−3, −1), *G*(0, 4) and *H*(3, −1) in the line $y = x$

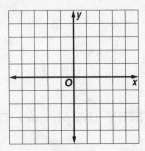

5. rectangle *QRST* with vertices *Q*(−3, 2), *R*(−1, 4), *S*(2, 1), and *T*(0, −1) in the *x*-axis

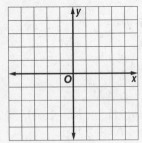

6. trapezoid *HIJK* with vertices *H*(−2, 5), *I*(2, 5), *J*(−4, −1), and *K*(−4, 3) in the *y*-axis

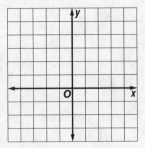

9-2 Skills Practice

Translations

Use the figure and the given translation vector. Then draw the translation of the figure along the translation vector.

1.

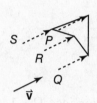

2.

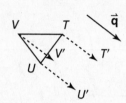

3.

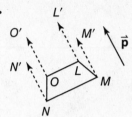

4.

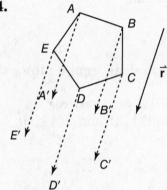

Graph each figure and its image along the given vector.

5. △JKL with vertices J(−4, −4), K(−2, −1), and L(2, −4); ⟨2, 5⟩

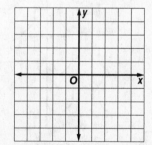

6. quadrilateral LMNP with vertices L(4, 2), M(4, −1), N(0, −1), and P(1, 4); ⟨−4, −3⟩

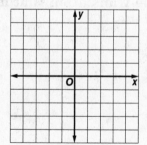

9-2 **Practice**

Translations

Use the figure and the given translation vector. Then draw the translation of the figure along the given translation vector.

1.

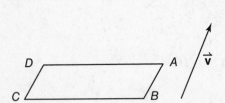

2.

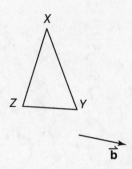

Graph each figure and its image along the given vector.

3. quadrilateral *TUWX* with vertices *T*(−1, 1), *U*(4, 2), *W*(1, 5), and *X*(−1, 3); ⟨−2, −4⟩

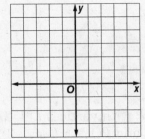

4. pentagon *DEFGH* with vertices *D*(−1, −2), *E*(2, −1), *F*(5, −2), *G*(4, −4), and *H*(1, −4); ⟨−1, 5⟩

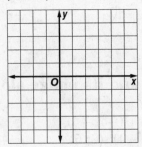

ANIMATION Find the translation that moves the figure on the coordinate plane.

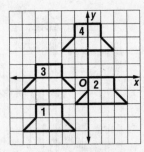

5. figure 1 → figure 2

6. figure 2 → figure 3

7. figure 3 → figure 4

9-3 Skills Practice

Rotations

Use a protractor and ruler to draw the specified rotation of each figure about point *K*.

1. 30°

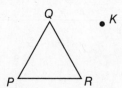

2. 150°

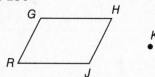

Graph each figure and its image after the specified rotation about the origin.

3. △*STU* has vertices *S*(2, −1), *T*(5, 1) and *U*(3, 3); 90°

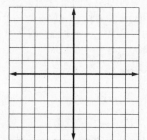

4. △*DEF* has vertices *D*(−4, 3), *E*(1, 2), and *F*(−3, −3); 180°

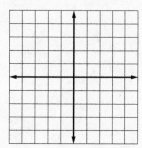

5. quadrilateral *WXYZ* has vertices *W*(−1, 8), *X*(0, 4), *Y*(−2, 1) and *Z*(−4, 3); 180°

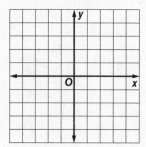

6. trapezoid *ABCD* has vertices *A*(9, 0), *B*(6, −7), *C*(3, −7) and *D*(0, 0); 270°

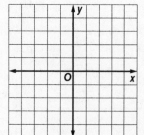

9-3 Practice

Rotations

Use a protractor and ruler to draw the specified rotation of each figure about point K.

1. 110°

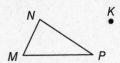

2. 280°

Graph each figure and its image after the specified rotation about the origin.

3. △PQR with vertices P(1, 3), Q(3, −2) and R(4, 2); 90°

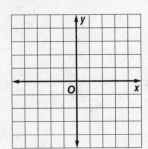

4. △ABC with vertices A(−4, 4), B(−2, −1) and C(2, −4); 270°

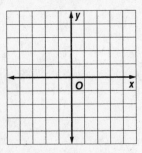

5. quadrilateral WXYZ with vertices W(1, 3), X(3, 1), Y(−6, 5) and Z(−5, 6); 180°

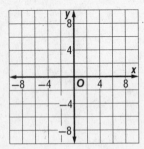

6. trapezoid FGHI with vertices F(8, 7), G(5, 8), H(−7, −2) and I(−3, −7); 90°

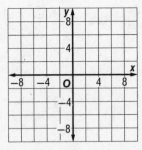

7. A ceiling fan has five equally spaced blades. Find the angle of rotation that maps one blade onto the adjacent blade.

9-4 Skills Practice

Compositions of Transformations

Figure $\overline{DE}$ has vertices $D(1, 3)$ and $E(3, -3)$. Graph $\overline{DE}$ and its image after the indicated transformation.

1. Translation: along $\langle 0, -2 \rangle$
Reflection: in x-axis

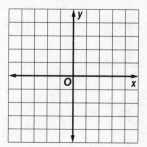

2. Translation: along $\langle 1, 2 \rangle$
Reflection: in y-axis

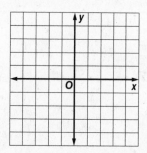

3. Translation: along $\langle -2, -1 \rangle$
Rotation: 270° about the origin

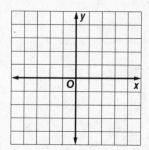

4. Reflection: in x-axis
Rotation: 90° about the origin

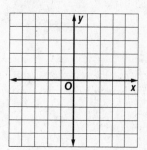

Copy and reflect figure F in line u and then line v. Then describe a single transformation that maps F into F''.

5.

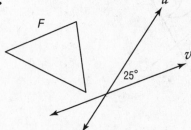

6.

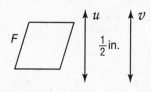

7.

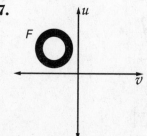

8.

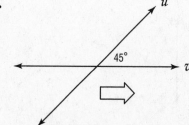

9-4 Practice

Composition of Transformations

Triangle *ABC* has vertices *A*(1, 3), *B*(−2, −1) and *C*(3, −2). Graph △*ABC* and its image after the indicated glide reflection.

1. Translation: along ⟨2, 0⟩
   Reflection: in *y*-axis

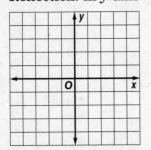

2. Translation: along ⟨−1, 1⟩
   Reflection: in *y* = *x*

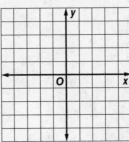

3. Translation: along ⟨−1, 2⟩
   Reflection: in *x* = *y*

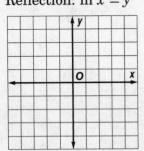

4. Translation: along ⟨0, 2⟩
   Reflection: in *y*-axis

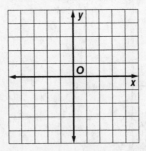

Copy and reflect figure *F* in line *x* and then line *y*. Then describe a single transformation that maps *F* into *F″*.

5.

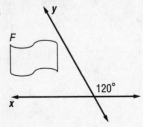

6.

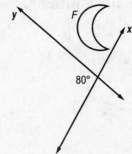

7.

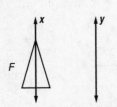

8.

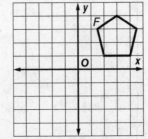

9-5 Skills Practice

Symmetry

State whether the figure appears to have line symmetry. Write *yes* or *no*. If so, draw all lines of symmetry and state their number.

1.

2.

3.

4.

State whether the figure has rotational symmetry. Write *yes* or *no*. If so, locate the center of symmetry, and state the order and magnitude of symmetry.

5.

6.

7.

8.

State whether the figure has *plane* symmetry, *axis* symmetry, *both*, or *neither*.

9.

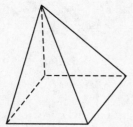

10.

9-5 Practice

Symmetry

State whether the figure has line symmetry. Write *yes* or *no*. If so, draw all lines of symmetry and state their number.

1.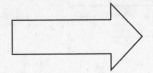

2.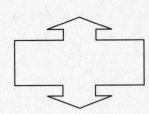

3.

State whether the figure has rotational symmetry. Write *yes* or *no*. If so, locate the center of symmetry and state the order and magnitude of symmetry.

4.

5.

6.

State whether the figure has *plane* symmetry, *axis* symmetry, *both*, or *neither*.

7.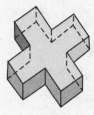

8.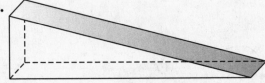

9. **STEAMBOATS** A paddle wheel on a steamboat is driven by a steam engine that rotates the paddles attached to the wheel to propel the boat through the water. If a paddle wheel consists of 18 evenly spaced paddles, identify the order and magnitude of its rotational symmetry.

9-6 Skills Practice

Dilations

Use the figure and point *C*. Then use a ruler to draw the image of the figure under a dilation with center *C* and the scale factor *r* indicated.

1. $r = 2$

2. $r = \frac{1}{4}$

Determine whether the dilation from Figure *K* to *K'* is an *enlargement* or a *reduction*. Then find the scale factor of the dilation and *x*.

3.

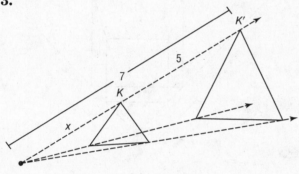

4.

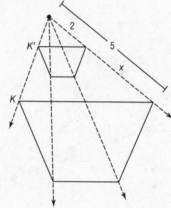

Find the image of each polygon with the given vertices after a dilation centered at the origin with the given scale factor.

5. $J(2, 4), K(4, 4), P(3, 2); r = 2$

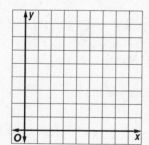

6. $D(-2, 0), G(0, 2), F(2, -2); r = 1.5$

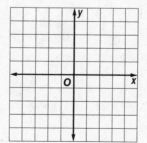

9-6 Practice

Dilations

Use a ruler to draw the image of the figure under a dilation with center C and the indicated scale factor r.

1. $r = \dfrac{3}{2}$

2. $r = \dfrac{2}{3}$

Determine whether the dilation from K to K' is an *enlargement* or a *reduction*. Then find the scale factor of the dilation and x.

3.

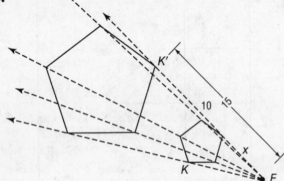

4.

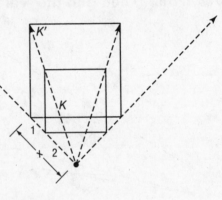

Graph the image of each polygon with the given vertices after a dilation centered at the origin with the given scale factor.

5. $A(1, 1)$, $C(2, 3)$, $D(4, 2)$, $E(3, 1)$; $r = 0.5$

6. $Q(-1, -1)$, $R(0, 2)$, $S(2, 1)$; $r = \dfrac{3}{2}$

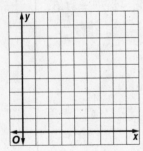

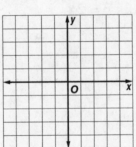

7. PHOTOGRAPHY Estebe enlarged a 4-inch by 6-inch photograph by a factor of $\dfrac{5}{2}$. What are the new dimensions of the photograph?

10-1 Skills Practice

Circles and Circumference

For Exercises 1–7, refer to ⊙*P*.

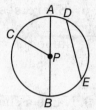

1. Name the circle.

2. Name a radius.

3. Name a chord.

4. Name a diameter.

5. Name a radius not drawn as part of a diameter.

6. Suppose the diameter of the circle is 16 centimeters. Find the radius.

7. If *PC* = 11 inches, find *AB*.

The diameters of ⊙*F* and ⊙*G* are 5 and 6 units, respectively. Find each measure.

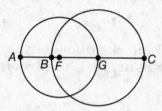

8. *BF*

9. *AB*

Find the diameter and radius of a circle with the given circumference. Round to the nearest hundredth.

10. *C* = 36 m

11. *C* = 17.2 ft

12. *C* = 81.3 cm

13. *C* = 5 yd

Find the exact circumference of each circle.

14.

3 cm

15.

8 ft

15 ft

10-1 Practice

Circles and Circumference

For Exercises 1–7, refer to ⊙L.

1. Name the circle.

2. Name a radius.

3. Name a chord.

4. Name a diameter.

5. Name a radius not drawn as part of a diameter.

6. Suppose the radius of the circle is 3.5 yards. Find the diameter.

7. If $RT = 19$ meters, find LW.

The diameters of ⊙L and ⊙M are 20 and 13 units, respectively, and $QR = 4$. Find each measure.

8. LQ

9. RM

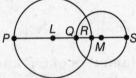

Find the diameter and radius of a circle with the given circumference. Round to the nearest hundredth.

10. $C = 21.2$ ft

11. $C = 5.9$ m

Find the exact circumference of each circle using the given inscribed or circumscribed polygon.

12.

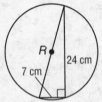

13.

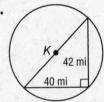

14. SUNDIALS Herman purchased a sundial to use as the centerpiece for a garden. The diameter of the sundial is 9.5 inches.

a. Find the radius of the sundial.

b. Find the circumference of the sundial to the nearest hundredth.

10-2 Skills Practice

Measuring Angles and Arcs

$\overline{AC}$ and $\overline{EB}$ are diameters of ⊙R. Identify each arc as a *major arc*, *minor arc*, or *semicircle* of the circle. Then find its measure.

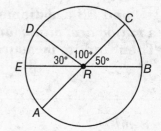

1. $m\widehat{EA}$

2. $m\widehat{CB}$

3. $m\widehat{DC}$

4. $m\widehat{DEB}$

5. $m\widehat{AB}$

6. $m\widehat{CDA}$

$\overline{PR}$ and $\overline{QT}$ are diameters of ⊙A. Find each measure.

7. $m\widehat{UPQ}$

8. $m\widehat{PQR}$

9. $m\widehat{UTS}$

10. $m\widehat{RS}$

11. $m\widehat{RSU}$

12. $m\widehat{STP}$

13. $m\widehat{PQS}$

14. $m\widehat{PRU}$

Use ⊙D to find the length of each arc. Round to the nearest hundredth.

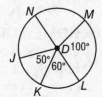

15. $\widehat{LM}$ if the radius is 5 inches

16. $\widehat{MN}$ if the diameter is 3 yards

17. $\widehat{KL}$ if JD = 7 centimeters

18. $\widehat{NJK}$ if NL = 12 feet

19. $\widehat{KLM}$ if DM = 9 millimeters

20. $\widehat{JK}$ if KD = 15 inches

10-2 Practice

Measuring Angles and Arcs

$\overline{AC}$ and $\overline{EB}$ are diameters of $\odot Q$. Identify each arc as a *major arc*, *minor arc*, or *semicircle* of the circle. Then find its measure.

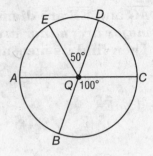

1. $m\widehat{AE}$

2. $m\widehat{AB}$

3. $m\widehat{EDC}$

4. $m\widehat{ADC}$

5. $m\widehat{ABC}$

6. $m\widehat{BC}$

$\overline{FH}$ and $\overline{EG}$ are diameters of $\odot P$. Find each measure.

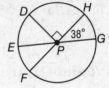

7. $m\widehat{EF}$

8. $m\widehat{DE}$

9. $m\widehat{FG}$

10. $m\widehat{DHG}$

11. $m\widehat{DFG}$

12. $m\widehat{DGE}$

Use $\odot Z$ to find each arc length. Round to the nearest hundredth.

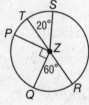

13. $\widehat{QPT}$, if $QZ = 10$ inches

14. $\widehat{QR}$, if $PZ = 12$ feet

15. $\widehat{PQR}$, if $TR = 15$ meters

16. $\widehat{QPS}$, if $ZQ = 7$ centimeters

17. **HOMEWORK** Refer to the table, which shows the number of hours students at Leland High School say they spend on homework each night.

Homework	
Less than 1 hour	8%
1–2 hours	29%
2–3 hours	58%
3–4 hours	3%
Over 4 hours	2%

   a. If you were to construct a circle graph of the data, how many degrees would be allotted to each category?

   b. Describe the arcs associated with each category.

10-3 Skills Practice

Arcs and Chords

ALGEBRA Find the value of x in each circle.

1.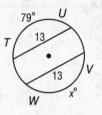

2.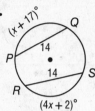

3.

4.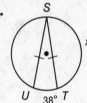

5.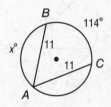

6.

In $\odot Y$ the radius is 34, $AB = 60$, and $m\widehat{AC} = 71$. Find each measure.

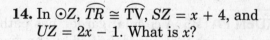

7. $m\widehat{BC}$

8. $m\widehat{AB}$

9. AD

10. BD

11. YD

12. DC

13. In $\odot U$, $VW = 20$ and $YZ = 5x$. What is x?

14. In $\odot Z$, $\widehat{TR} \cong \widehat{TV}$, $SZ = x + 4$, and $UZ = 2x - 1$. What is x?

10-3 Practice

Arcs and Chords

ALGEBRA Find the value of x in each circle.

1.
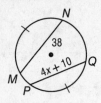

2.
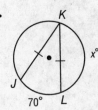

3.
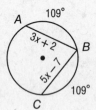

4. $\odot R \cong \odot S$
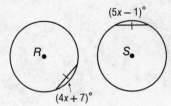

The radius of $\odot N$ is 18, $NK = 9$, and $m\overset{\frown}{DE} = 120$. Find each measure.

5. $m\overset{\frown}{GE}$

6. $m\angle HNE$

7. $m\angle HEN$

8. HN

9. In $\odot P$, $QR = 7x - 20$ and $TS = 3x$. What is x?

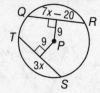

10. In $\odot K$, $\overline{JL} \cong \overline{LM}$, $KN = 3x - 2$, and $KP = 2x + 1$. What is x?

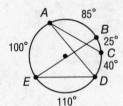

11. **GARDEN PATHS** A circular garden has paths around its edge that are identified by the given arc measures. It also has four straight paths, identified by segments $\overline{AC}$, $\overline{AD}$, $\overline{BE}$, and $\overline{DE}$, that cut through the garden's interior. Which two straight paths have the same length?

10-4 Skills Practice

Inscribed Angles

Find each measure.

1. $m\widehat{XY}$

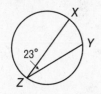

2. $m\angle E$

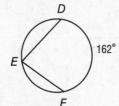

3. $m\angle R$

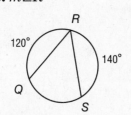

4. $m\widehat{MP}$

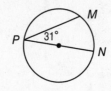

ALGEBRA Find each measure.

5. $m\angle N$

7. $m\angle L$

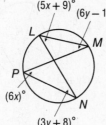

6. $m\angle C$

8. $m\angle A$

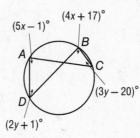

9. $m\angle J$

11. $m\angle K$

10. $m\angle S$

12. $m\angle R$

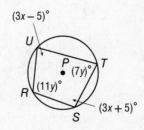

10-4 Practice

Inscribed Angles

Find each measure.

1. $m\widehat{AB}$

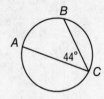

2. $m\angle X$

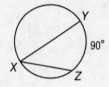

3. $m\widehat{JK}$

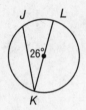

4. $m\angle Q$

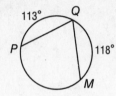

ALGEBRA Find each measure.

5. $m\angle W$

6. $m\angle Y$

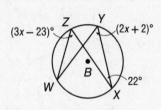

7. $m\angle A$

8. $m\angle D$

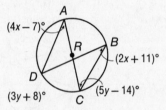

ALGEBRA Find each measure.

9. $m\angle A$

10. $m\angle C$

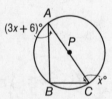

11. $m\angle G$

12. $m\angle H$

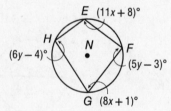

13. PROBABILITY In $\odot V$, point C is randomly located so that it does not coincide with points R or S. If $m\widehat{RS} = 140$, what is the probability that $m\angle RCS = 70$?

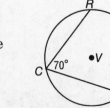

10-5 Skills Practice

Tangents

Determine whether each segment is tangent to the given circle. Justify your answer.

1. $\overline{HI}$

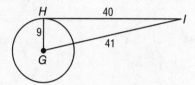

2. $\overline{AB}$

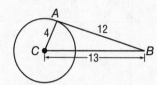

Find x. Assume that segments that appear to be tangent are tangent. Round to the nearest tenth if necessary.

3.

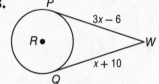

4.

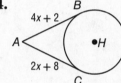

5.

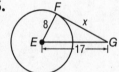

6.

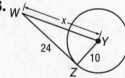

For each figure, find x. Then find the perimeter.

7.

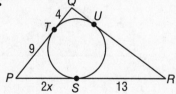

8.

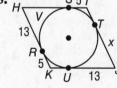

10-5 Practice

Tangents

Determine whether each segment is tangent to the given circle. Justify your answer.

1. $\overline{MP}$

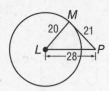

2. $\overline{QR}$

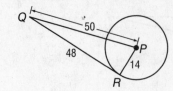

Find x. Assume that segments that appear to be tangent are tangent. Round to the nearest tenth if necessary.

3.

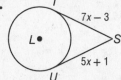

4.

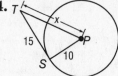

For each figure, find x. Then find the perimeter.

5.

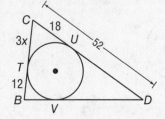

6.

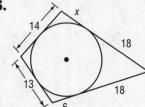

7. CLOCKS The design shown in the figure is that of a circular clock face inscribed in a triangular base. AF and FC are equal.

a. Find AB.

b. Find the perimeter of the clock.

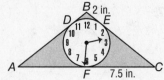

10-6 Skills Practice

Secants, Tangents, and Angle Measures

Find each measure. Assume that segment that appear to be tangent are tangent.

1. $m\angle 1$

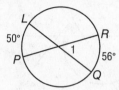

2. $m\angle 2$

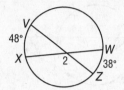

3. $m\angle 3$

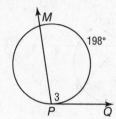

4. $m\angle 4$

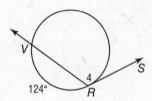

5. $m\angle 5$

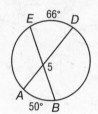

6. $m\angle 6$

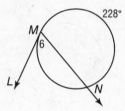

7. $m\angle R$

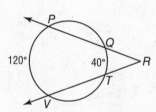

8. $m\angle K$

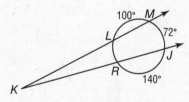

9. $m\angle U$

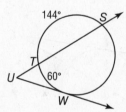

10. $m\angle S$

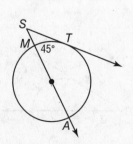

11. $m\widehat{DPA}$

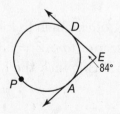

12. $m\widehat{LJ}$

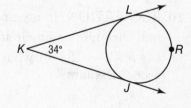

10-6 Practice

Secants, Tangents, and Angle Measures

Find each measure. Assume that any segments that appear to be tangent are tangent.

1. $m\angle 1$

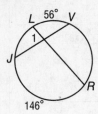

2. $m\angle 2$

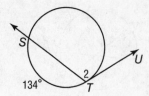

3. $m\angle 3$

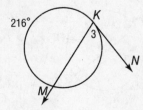

4. $m\angle R$

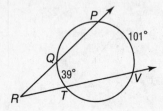

5. $m\widehat{GJ}$

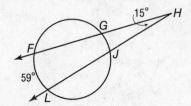

6. $m\angle R$

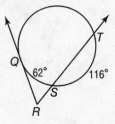

7. $m\angle Y$

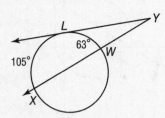

8. $m\widehat{CE}$

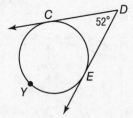

9. $m\widehat{YAB}$

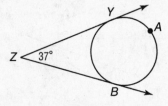

10. RECREATION In a game of kickball, Rickie has to kick the ball through a semicircular goal to score. If $m\widehat{XZ} = 58$ and the $m\widehat{XY} = 122$, at what angle must Rickie kick the ball to score? Explain.

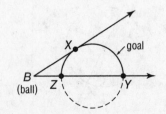

10-7 Skills Practice

Special Segments in a Circle

Find x to the nearest tenth if necessary. Assume that segments that appear to be tangent are tangent.

1.

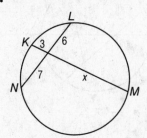

2.

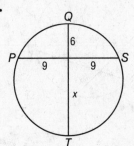

3.

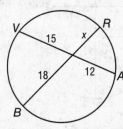

4.

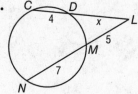

5.

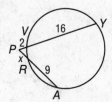

6.

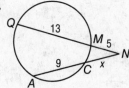

7.

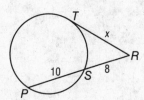

8.

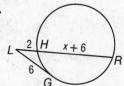

9.

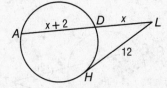

10-7 Practice

Special Segments in a Circle

Find *x*. Assume that segments that appear to be tangent are tangent. Round to the nearest tenth if necessary.

1.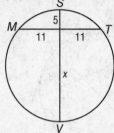

2.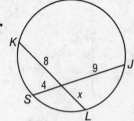

3.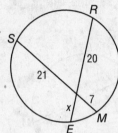

4.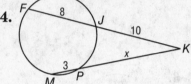

5.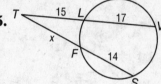

6.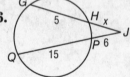

7.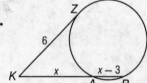

8.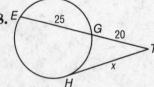

9.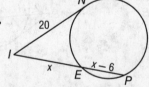

10. **CONSTRUCTION** An arch over an apartment entrance is 3 feet high and 9 feet wide. Find the radius of the circle containing the arc of the arch.

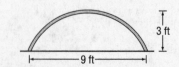

10-8 Skills Practice

Equations of Circles

Write the equation of each circle.

1. center at origin, radius 6

2. center at (0, 0), radius 2

3. center at (4, 3), radius 9

4. center at (7, 1), diameter 24

5. center at (−4, −1), passes through (−2, 3)

6. center at (5, −2), passes through (4, 0)

7.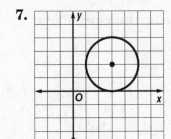

8.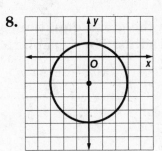

For each circle with the given equation, state the coordinates of the center and the measure of the radius. Then graph the equation.

9. $x^2 + y^2 = 16$

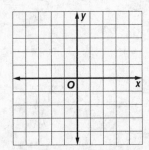

10. $(x - 1)^2 + (y - 4)^2 = 9$

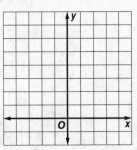

Write an equation of a circle that contains each set of points. Then graph the circle.

11. $A(-2, 3)$, $B(1, 0)$, $C(4, 3)$

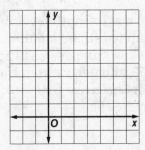

12. $F(3, 0)$, $G(5, -2)$, $H(1, -2)$

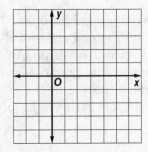

10-8 Practice

Equations of Circles

Write the equation of each circle.

1. center at origin, radius 7

2. center at (0, 0), diameter 18

3. center at (−7, 11), radius 8

4. center at (12, −9), diameter 22

5. center at (−1, 8), passes through (9, 3)

6. center at (−3, −3), passes through (−2, 3)

For each circle with the given equation, state the coordinates of the center and the measure of the radius. Then graph the equation.

7. $x^2 + y^2 = 4$

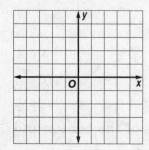

8. $(x + 3)^2 + (y − 3)^2 = 9$

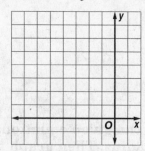

Write an equation of a circle that contains each set of points. Then graph the circle.

9. $A(−2, 2), B(2, −2), C(6, 2)$

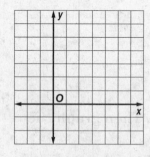

10. $R(5, 0), S(−5, 0), T(0, −5)$

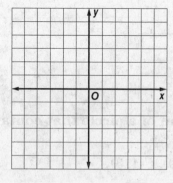

11. EARTHQUAKES When an earthquake strikes, it releases seismic waves that travel in concentric circles from the epicenter of the earthquake. Seismograph stations monitor seismic activity and record the intensity and duration of earthquakes. Suppose a station determines that the epicenter of an earthquake is located about 50 kilometers from the station. If the station is located at the origin, write an equation for the circle that represents one of the concentric circles of seismic waves of the earthquake.

11-1 Skills Practice

Areas of Parallelograms and Triangles

Find the perimeter and area of each parallelogram or triangle. Round to the nearest tenth if necessary.

1.
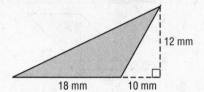
18 mm   10 mm   12 mm

2.
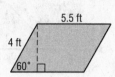
5.5 ft
4 ft
60°

3.
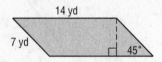
14 yd
7 yd
45°

4.
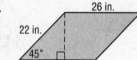
26 in.
22 in.
45°

5.

3.4 m

6.
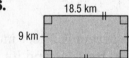
18.5 km
9 km

7.
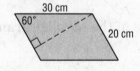
30 cm
60°
20 cm

8.
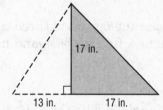
17 in.
13 in.   17 in.

9. The height of a parallelogram is 10 feet more than its base. If the area of the parallelogram is 1200 square feet, find its base and height.

10. The base of a triangle is one half of its height. If the area of the triangle is 196 square millimeters, find its base and height.

11-1 Practice

Areas of Parallelograms and Triangles

Find the perimeter and area of each parallelogram or triangle. Round to the nearest tenth if necessary.

1.
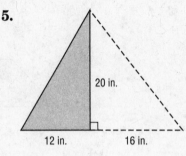

2.

3.
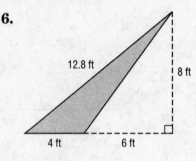

4.

5.

6.

7. The height of a parallelogram is 5 feet more than its base. If the area of the parallelogram is 204 square feet, find its base and height.

8. The height of a parallelogram is three times its base. If the area of the parallelogram is 972 square inches, find its base and height.

9. The base of a triangle is four times its height. If the area of the triangle is 242 square millimeters, find its base and height.

10. **FRAMING** A rectangular poster measures 42 inches by 26 inches. A frame shop fitted the poster with a half-inch mat border.

   a. Find the area of the poster.

   b. Find the area of the mat border.

   c. Suppose the wall is marked where the poster will hang. The marked area includes an additional 12-inch space around the poster and frame. Find the total wall area that has been marked for the poster.

11-2 Skills Practice

Areas of Trapezoids, Rhombi, and Kites

Find the area of each trapezoid, rhombus, or kite.

1.
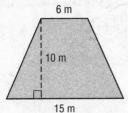
6 m
10 m
15 m

2.
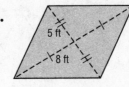
12 mm
14 mm

3.
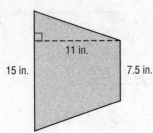
11 in.
15 in.   7.5 in.

4.
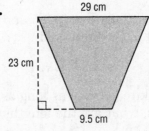
5 ft
8 ft

5.
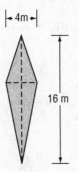
4m
16 m

6.
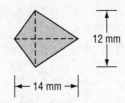
29 cm
23 cm
9.5 cm

ALGEBRA Find each missing length.

7. A trapezoid has base lengths of 6 and 15 centimeters with an area of 136.5 square centimeters. What is the height of the trapezoid?

8. One diagonal of a kite is four times as long as the other diagonal. If the area of the kite is 72 square meters, what are the lengths of the diagonals?

9. A trapezoid has a height of 24 meters, a base of 4 meters, and an area of 264 square meters. What is the length of the other base?

11-2 Practice

Areas of Trapezoids, Rhombi, and Kites

Find the area of each trapezoid, rhombus, or kite.

1.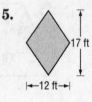
31 m
5 m
16 m

2.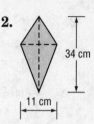
34 cm
11 cm

3.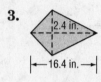
2.4 in.
16.4 in.

4.
6.5 ft
8 ft
21.5 ft

5.
17 ft
12 ft

6.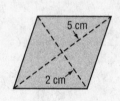
5 cm
2 cm

ALGEBRA Find each missing length.

7. A trapezoid has base lengths of 19.5 and 24.5 centimeters with an area of 154 cm². What is the height of the trapezoid?

8. One diagonal of a kite is twice as long as the other diagonal. If the area of the kite is 400 square meters, what are the lengths of the diagonals?

9. A trapezoid has a height of 40 inches, a base of 15 inches, and an area of 2400 square inches. What is the length of the other base?

10. DESIGN Mr. Hagarty used 16 congruent rhombi-shaped tiles to design the midsection of the backsplash area above a kitchen sink. The length of the design is 27 inches and the total area is 108 square inches.

a. Find the area of one rhombus.

b. Find the length of each diagonal.

11-3 Skills Practice

Areas of Circles and Sectors

Find the area of each circle.

1.

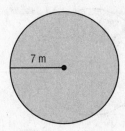

7 m

2.

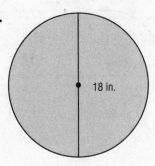

18 in.

3.

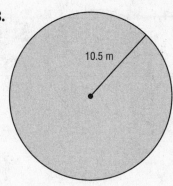

10.5 m

Find the indicated measure. Round to the nearest tenth.

4. The area of a circle is 132.7 square centimeters. Find the diameter.

5. Find the diameter of a circle with an area of 1134.1 square millimeters.

6. The area of a circle is 706.9 square inches. Find the radius.

7. Find the radius of a circle with an area of 2827.4 square feet.

Find the area of each shaded sector. Round to the nearest tenth.

8.

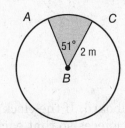

A   C
51° 2 m
B

9.

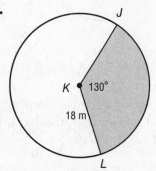

J
K 130°
18 m
L

10.

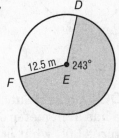

D
12.5 m 243°
F   E

11. GAMES Jason wants to make a spinner for a new board game
he invented. The spinner is a circle divided into 8 congruent
pieces, what is the area of each piece to the nearest tenth?

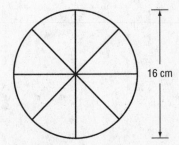

16 cm

11-3 Practice

Areas of Circles and Sectors

Find the area of each circle. Round to the nearest tenth.

1.

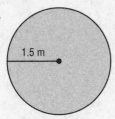

2.

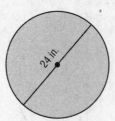

3.

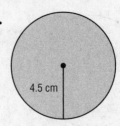

Find the indicated measure. Round to the nearest tenth.

4. The area of a circle is 3.14 square centimeters. Find the diameter.

5. Find the diameter of a circle with an area of 855.3 square millimeters.

6. The area of a circle is 201.1 square inches. Find the radius.

7. Find the radius of a circle with an area of 2290.2 square feet.

Find the area of each shaded sector. Round to the nearest tenth.

8.

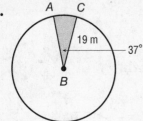

9.

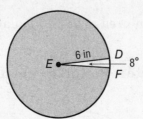

10.

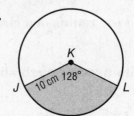

11. **CLOCK** Sadie wants to draw a clock face on a circular piece of cardboard. If the clock face has a diameter of 20 centimeters and is divided into congruent pieces so that each sector is 30°, what is the area of each piece?

11-4　Skills Practice

Areas of Regular Polygons and Composite Figures

Find the area of each regular polygon. Round to the nearest tenth.

1.

8 m

2.

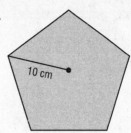

10 cm

3.

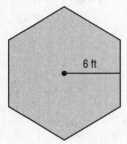

6 ft

4.

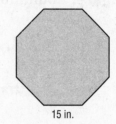

15 in.

Find the area of each figure. Round to the nearest tenth if necessary.

5.

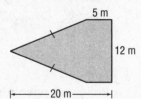

5 m

12 m

20 m

6.

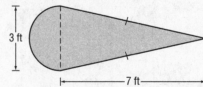

3 ft

7 ft

7.

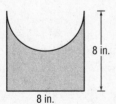

8 in.

8 in.

8.

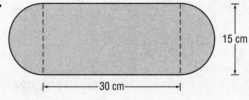

15 cm

30 cm

11-4 Practice

Areas of Regular Polygons and Composite Figures

Find the area of each regular polygon. Round to the nearest tenth.

1.

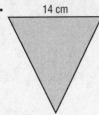

14 cm

2.

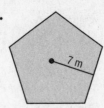

7 m

Find the area of each figure. Round to the nearest tenth if necessary.

3.

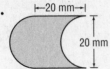

|←20 mm→|

20 mm

4.

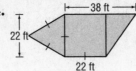

38 ft

22 ft

22 ft

5.

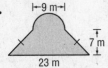

|←9 m→|

7 m

23 m

6.

|←20 in.→|

13 in.   30 in.

13 in.

7. LANDSCAPING One of the displays at a botanical garden is a koi pond with a walkway around it. The figure shows the dimensions of the pond and the walkway.

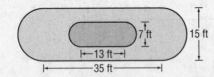

7 ft

15 ft

|←13 ft→|

|←————35 ft————→|

a. Find the area of the pond to the nearest tenth.

b. Find the area of the walkway to the nearest tenth.

11-5 Skills Practice

Areas of Similar Figures

For each pair of similar figures, find the area of the shaded figure.

1.

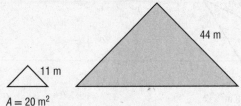
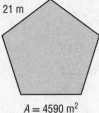

$A = 20 \text{ m}^2$

2.

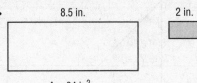

8.5 in.   2 in.

$A = 34 \text{ in}^2$

For each pair of similar figures, use the given areas to find the scale factor from the unshaded to the shaded figure. Then find x.

3.

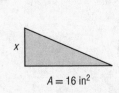

21 m x

$A = 4590 \text{ m}^2$ $A = 510 \text{ m}^2$

4.

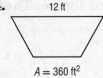

12 ft x

$A = 360 \text{ ft}^2$ $A = 10 \text{ ft}^2$

5.

x

9.5 in.

$A = 16 \text{ in}^2$ $A = 71 \text{ in}^2$

6.

14 ft x

$A = 588 \text{ ft}^2$ $A = 272 \text{ ft}^2$

7. SCIENCE PROJECT Matt has two posters for his science project. Each poster is a rectangle. The length of the larger poster is 11 inches. The length of the smaller poster is 6 inches. What is the area of the smaller poster if the larger poster is 93.5 square inches?

11-5 Practice

Areas of Similar Figures

For each pair of similar figures, find the area of the shaded figure.

1.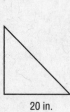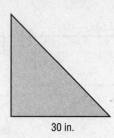

20 in.   30 in.

$A = 200 \text{ in}^2$

2.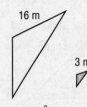

16 m

3 m

$A = 38 \text{ m}^2$

For each pair of similar figures, use the given areas to find the scale factor from the unshaded to the shaded figure. Then find x.

3.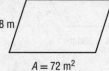

x m

8 m

$A = 50 \text{ m}^2$ $A = 72 \text{ m}^2$

4.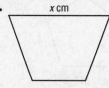

x cm   7 cm

$A = 30 \text{ cm}^2$

$A = 70 \text{ cm}^2$

5.

x ft   8 ft

$A = 16 \text{ ft}^2$

$A = 64 \text{ ft}^2$

6.

9 cm x cm

$A = 13 \text{ cm}^2$

$A = 39 \text{ cm}^2$

7. ARCHERY A target consists of two concentric similar octagons. The outside octagon has a side length of 2 feet and an area of 19.28 square feet. If the inside octagon has a side length of 1.5 feet, what is the area of the inside octagon?

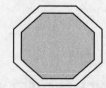

12-1 Skills Practice

Representations of Three-Dimensional Figures

Use isometric dot paper to sketch each prism.

1. cube 2 units on each edge

2. rectangular prism 2 units high, 5 units long, and 2 units wide

Use isometric dot paper and each orthographic drawing to sketch a solid.

3.

top view   left view   front view   right view

4.

top view   left view   front view   right view

Describe each cross section.

5.

6.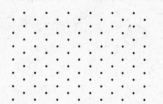

7.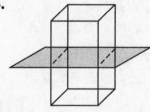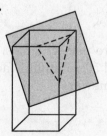

8.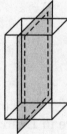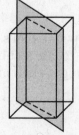

12-1 Practice

Representations of Three-Dimensional Figures

Use isometric dot paper to sketch each prism.

1. rectangular prism 3 units high, 3 units long, and 2 units wide

2. triangular prism 3 units high, whose bases are right triangles with legs 2 units and 4 units long

Use isometric dot paper and each orthographic drawing to sketch a solid.

3.

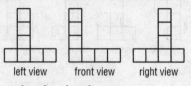

top view   left view   front view   right view

4.

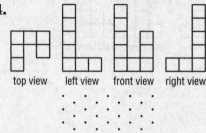

top view   left view   front view   right view

Sketch the cross section from a vertical slice of each figure.

5.

6.

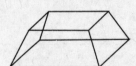

7. SPHERES Consider the sphere in Exercise 5. Based on the cross section resulting from a horizontal and a vertical slice of the sphere, make a conjecture about all spherical cross sections.

8. MINERALS Pyrite, also known as fool's gold, can form crystals that are perfect cubes. Suppose a gemologist wants to cut a cube of pyrite to get a square and a rectanglar face. What cuts should be made to get each of the shapes? Illustrate your answers.

12-2 Skills Practice

Surface Areas of Prisms and Cylinders

Find the lateral area and surface area of each prism. Round to the nearest tenth if necessary.

1.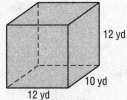

12 yd
10 yd
12 yd

2.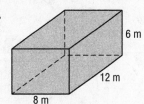

6 m
12 m
8 m

3.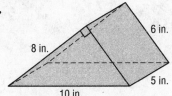

6 in.
8 in.
5 in.
10 in.

4.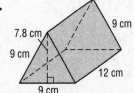

7.8 cm
9 cm
9 cm
9 cm
12 cm

Find the lateral area and surface area of each cylinder. Round to the nearest tenth.

5.

10 in.
12 in.

6.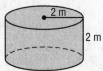

2 m
2 m

7.

3 yd
2 yd

8.

8 in.
12 in.

12-2 Practice

Surface Areas of Prisms

Find the lateral and surface area of each prism. Round to the nearest tenth if necessary.

1.

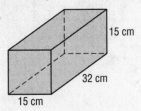

15 cm
32 cm
15 cm

2.

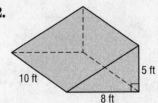

10 ft
5 ft
8 ft

3.

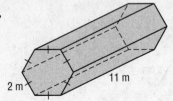

2 m
11 m

4.

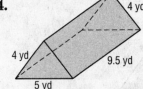

4 yd
4 yd
9.5 yd
5 yd

Find the lateral area and surface area of each cylinder. Round to the nearest tenth.

5.

5 ft
7 ft

6.

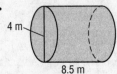

4 m
8.5 m

7.

19 in.
17 in.

8.

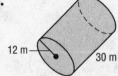

12 m
30 m

12-3 Skills Practice

Surface Areas of Pyramids and Cones

Find the lateral area and surface area of each regular pyramid. Round to the nearest tenth if necessary.

1.

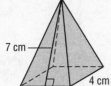

7 cm

4 cm

2.

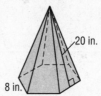

20 in.

8 in.

3.

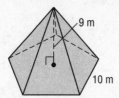

9 m

10 m

4.

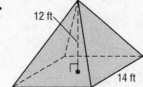

12 ft

14 ft

Find the lateral area and surface area of each cone. Round to the nearest tenth.

5.

5 m

14 m

6.

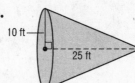

10 ft

25 ft

7.

21 in.

8 in.

8.

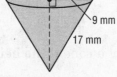

9 mm

17 mm

12-3 Practice

Surface Areas of Pyramids and Cones

Find the lateral area and surface area of each regular pyramid. Round to the nearest tenth if necessary.

1.

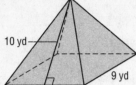

10 yd

9 yd

2.

12 m

7 m

3.

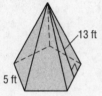

13 ft

5 ft

4.

8 cm

2.5 cm

Find the lateral area and surface area of each cone. Round to the nearest tenth if necessary.

5.

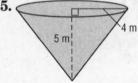

5 m

4 m

6.

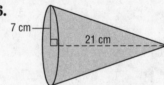

7 cm

21 cm

7. Find the surface area of a cone if the height is 14 centimeters and the slant height is 16.4 centimeters.

8. Find the surface area of a cone if the height is 12 inches and the diameter is 27 inches.

9. **GAZEBOS** The roof of a gazebo is a regular octagonal pyramid. If the base of the pyramid has sides of 0.5 meters and the slant height of the roof is 1.9 meters, find the area of the roof.

10. **HATS** Cuong bought a conical hat on a recent trip to central Vietnam. The basic frame of the hat is 16 hoops of bamboo that gradually diminish in size. The hat is covered in palm leaves. If the hat has a diameter of 50 centimeters and a slant height of 32 centimeters, what is the lateral area of the conical hat?

12-4 Skills Practice

Volumes of Prisms and Cylinders

Find the volume of each prism or cylinder. Round to the nearest tenth if necessary.

1.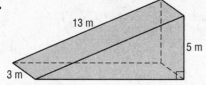
8 cm
16 cm
18 cm

2.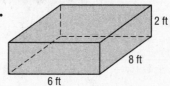
2 ft
8 ft
6 ft

3.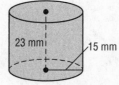
13 m
5 m
3 m

4.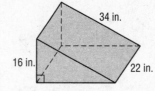
34 in.
16 in.
22 in.

5.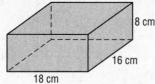
23 mm
15 mm

6.
6 yd
10 yd

Find the volume of each oblique prism or cylinder. Round to the nearest tenth if necessary.

7.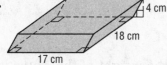
4 cm
18 cm
17 cm

8.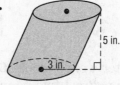
5 in.
3 in.

12-4   Practice

Volumes of Prisms and Cylinders

Find the volume of each prism or cylinder. Round to the nearest tenth if necessary.

1.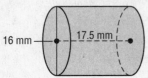
26 m
17 m
10 m

2.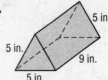
5 in.
5 in.
9 in.
5 in.

3.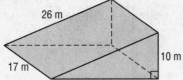
16 mm
17.5 mm

4.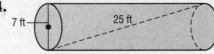
7 ft
25 ft

5.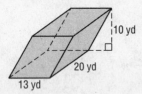
10 yd
20 yd
13 yd

6.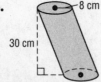
8 cm
30 cm

7. AQUARIUM Mr. Gutierrez purchased a cylindrical aquarium for his office. The aquarium has a height of $25\frac{1}{2}$ inches and a radius of 21 inches.

   a. What is the volume of the aquarium in cubic feet?

   b. If there are 7.48 gallons in a cubic foot, how many gallons of water does the aquarium hold?

   c. If a cubic foot of water weighs about 62.4 pounds, what is the weight of the water in the aquarium to the nearest five pounds?

12-5 Skills Practice

Volumes of Pyramids and Cones

Find the volume of each pyramid or cone. Round to the nearest tenth if necessary.

1.
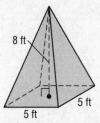
8 ft
5 ft
5 ft
5 ft

2.
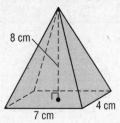
8 cm
7 cm
4 cm

3.
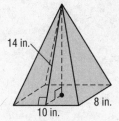
14 in.
10 in.
8 in.

4.
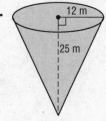
12 m
25 m

5.
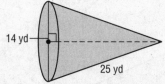
14 yd
25 yd

6.
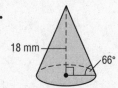
18 mm
66°

Find the volume of each oblique pyramid or cone. Round to the nearest tenth if necessary.

7.
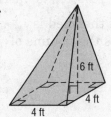
6 ft
4 ft
4 ft
4 ft

8.
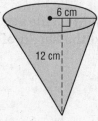
6 cm
12 cm

12-5 **Practice**

Volumes of Pyramids and Cones

Find the volume of each pyramid or cone. Round to the nearest tenth if necessary.

1.

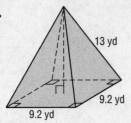

13 yd
9.2 yd
9.2 yd

2.

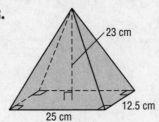

23 cm
12.5 cm
25 cm

3.

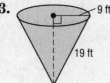

9 ft
19 ft

4.

12 mm
52°

5.

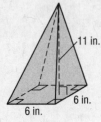

11 in.
6 in.
6 in.

6.

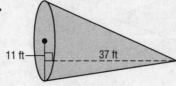

11 ft
37 ft

7. **CONSTRUCTION** Mr. Ganty built a conical storage shed. The base of the shed is 4 meters in diameter and the height of the shed is 3.8 meters. What is the volume of the shed?

8. **HISTORY** The start of the pyramid age began with King Zoser's pyramid, erected in the 27th century B.C. In its original state, it stood 62 meters high with a rectangular base that measured 140 meters by 118 meters. Find the volume of the original pyramid.

12-6 Skills Practice

Surface Areas and Volumes of Spheres

Find the surface area of each sphere or hemisphere. Round to the nearest tenth.

1.
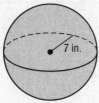

2.
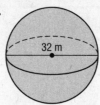

3. hemisphere: radius of great circle = 8 yd

4. sphere: area of great circle ≈ 28.6 in²

Find the volume of each sphere or hemisphere. Round to the nearest tenth.

5.
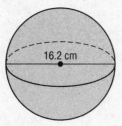

6.
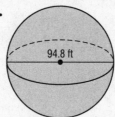

7. hemisphere: diameter = 48 yd

8. sphere: circumference of great circle ≈ 26 m

9. sphere: diameter = 10 in.

12-6 Practice

Surface Areas and Volumes of Spheres

Find the surface area of each sphere or hemisphere. Round to the nearest tenth.

1.

6.5 cm

2.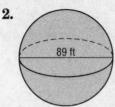

89 ft

3. hemisphere: radius of great circle = 8.4 in.

4. sphere: area of great circle ≈ 29.8 m²

Find the volume of each sphere or hemisphere. Round to the nearest tenth.

5.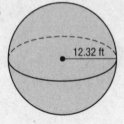

12.32 ft

6.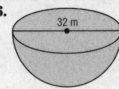

32 m

7. hemisphere: diameter = 18 mm

8. sphere: circumference ≈ 36 yd

9. sphere: radius = 12.4 in.

12-7 Skills Practice

Spherical Geometry

Name two lines containing point *K*, a segment containing point *T*, and a triangle in each of the following spheres.

1.
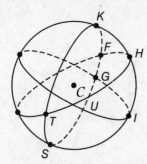

2.
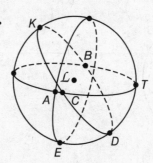

Determine whether figure *u* on each of the spheres shown is a line in spherical geometry.

3.
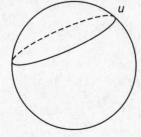

4.

basketball

Tell whether the following postulate or property of plane Euclidean geometry has a corresponding statement in spherical geometry. If so, write the corresponding statement. If not, explain your reasoning.

5. If two lines form vertical angles, then the angles are equal in measure.

6. If two lines meet a third line at the same angle, those lines are parallel.

7. Two lines meet at two 90° angles or they meet at angles whose sum is 180°.

8. Three non-parallel lines divide the plane into 7 separate parts.

12-7 Practice

Spherical Geometry

Name two lines containing point *K*, a segment containing point *T*, and a triangle in each of the following spheres.

1.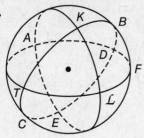

2.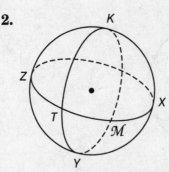

Determine whether figure *u* on each of the spheres shown is a line in spherical geometry.

3.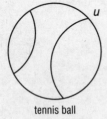

tennis ball

4.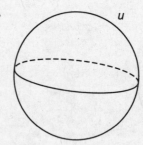

Tell whether the following postulate or property of plane Euclidean geometry has a corresponding statement in spherical geometry. If so, write the corresponding statement. If not, explain your reasoning.

5. A triangle can have at most one obtuse angle.

6. The sum of the angles of a triangle is 180°.

7. Given a line and a point not on the line, there is exactly one line that goes through the point and is perpendicular to the line.

8. All equilateral triangles are similar.

9. **AIRPLANES** When flying an airplane from New York to Seattle, what is the shortest route: flying directly west, or flying north across Canada? Explain.

12-8 Skills Practice

Congruent and Similar Solids

Determine whether each pair of solids is *similar*, *congruent*, or *neither*. If the solids are similar, state the scale factor.

1.

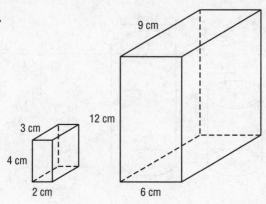

2.

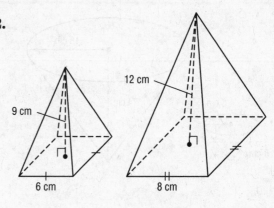

3.

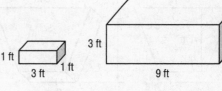

4.

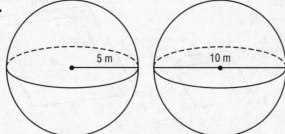

5. Two similar pyramids have heights of 4 inches and 7 inches What is the ratio of the volume of the small pyramid to the volume of the large pyramid?

6. Two similar cylinders have surface areas of 40π square feet and 90π square feet. What is the ratio of the height of the large cylinder to the height of the small cylinder?

7. COOKING Two stockpots are similar cylinders. The smaller stockpot has a height of 10 inches and a radius of 2.5 inches. The larger stockpot has a height of 16 inches. What is the volume of the larger stockpot? Round to the nearest tenth.

12-8 Practice

Congruent and Similar Solids

Determine whether the pair of solids is *similar*, *congruent*, or *neither*. If the solids are similar, state the scale factor.

1.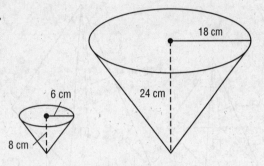

2.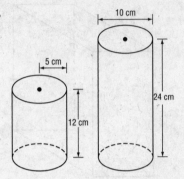

3.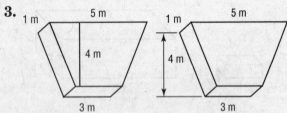

4.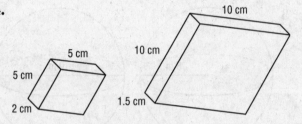

5. Two cubes have surface areas of 72 square feet and 98 square feet. What is the ratio of the volume of the small cube to the volume of the large cube?

6. Two similar ice cream cones are made of a half sphere on top and a cone on bottom. They have radii of 1 inch and 1.75 inches respectively. What is the ratio of the volume of the small ice cream cone to the volume of the large ice cream cone? Round to the nearest tenth.

7. **ARCITHECTURE** Architects make scale models of buildings to present their ideas to clients. If an architect wants to make a 1:50 scale model of a 4000 square foot house, how many square feet will the model have?

13-1 Skills Practice

Representing Sample Spaces

Represent the sample space for each experiment by making an organized list, a table, and a tree diagram.

1. Michelle could take a summer job in California or Arizona at a hotel or a bed-and-breakfast.

2. Jeremy could go to baseball or soccer camp as a counselor or an assistant director.

3. Brad could buy his mom a $25 or $50 gift card for a spa or a housecleaning service.

Find the number of possible outcomes for each situation.

4. Marie's family is buying a house. They must choose one from each category.

House Plans	Number of Choices
Subdivision location	4
Floor plans	5
Garage size	2
Front yard landscape package	3
Backyard pool package	3

5. Mr. Thomson is choosing his cable TV. He must choose one from each category.

Cable TV Plans	Number of Choices
Channel packages	16
DVR system	3
Contract length	3
Service contract	2
Include phone	2
Include Internet	2

6. Valentine gift sets come with a choice of 4 different teddy bears, 8 types of candy, 5 balloon designs, and 3 colors of roses.

7. Joni wears a school uniform that consists of a skirt or pants, a white shirt, a blue jacket or sweater, white socks and black shoes. She has 3 pairs of pants, 3 skirts, 6 shirts, 2 jackets, 2 sweaters, 6 pairs of socks and 3 pairs of black shoes.

13-1 Practice

Representing Sample Spaces

Represent the sample space for each experiment by making an organized list, a table, and a tree diagram.

1. Tavya can spend the summer with her cousins or her grandparents at the lake or at the beach.

2. Jordan can write his final essay in class or at home on a scientific or an historical topic.

3. Julio can join the Air Force or the Army before or after college.

Find the number of possible outcomes for each situation.

4. Josh is making a stuffed animal.

Animal Options	Number of Choices
Animals	10
Type of stuffing	3
Sound effect	5
Eye color	3
Outfit	20

5. Kelley is buying an ice cream cone. Assume one of each category is ordered.

Ice Cream	Number of Choices
Type of cone	3
Flavors	20
Cookie toppings	4
Candy toppings	8

6. Movie-themed gift baskets come with a choice of one of each of the following: 4 flavors of popcorn, 4 different DVDs, 4 types of drinks, and 8 different kinds of candy.

7. **INTERNSHIP** Jack is choosing an internship program that could take place in 3 different months, in 4 different departments of 3 different firms. Jack is only available to complete his internship in July. How many different outcomes are there for Jack's internship?

13-2 Skills Practice

Probabilities With Permutations and Combinations

1. **DISPLAY** The Art Club is displaying the students' works in the main hallway. In a row of 12 randomly ordered paintings, what is the probability that Tim's and Abby's paintings are in the 6th and 7th positions?

2. **LINE UP** When the 18 French class students randomly line up for a fire drill, what is the probability that Amy is first and Zach is last in line?

3. **TRY-OUTS** Ten students made call-backs for the three lead roles in the school play. What is the probability Sarah, Maria, and Jimenez will be chosen for the leads?

4. **SECURITY** Parking stickers contain randomly generated numbers with 5-digits ranging from 1 to 9. No digits are repeated. What is the probability that a randomly generated number is 54321?

5. **MEETING** Micah is arranging 15 chairs in a circle for an ice breaker game for the first club meeting. If people choose their seats randomly, what is the probability Micah sits in the seat closest to the door?

6. **MERRY-GO-ROUND** The mall has a merry-go-round with 12 horses on the outside ring. If 12 people randomly choose those horses, what is the probability they are seated in alphabetical order?

7. **PROMOTION** Tony is promoting his band's first concert. He contacts 10 local radio stations. If 4 of them agree to interview him on the air, what is the probability they are the top 4 stations in the area?

8. **TALENT SHOW** The Sign Language Club is choosing 10 of its 15 members to perform at the school talent show. What is the probability that the 10 people chosen are the 10 seniors in the club?

13-2 Practice

Probability with Permutations and Combinations

1. **FORMAL DINING** You are handed 5 pieces of silverware for the formal setting shown. If you guess their placement at random, what is the probability that the knife and spoon are placed correctly?

2. **GOLF** The standings list after the first day of a 3-day tournament is shown below. What is the probability that Wyatt, Gabe, and Isaac will all finish in the top 3?

DAY 1 STANDINGS	
MCAFEE, DAVID	−3
FORD, GABE	−2
STANDISH, TRISTAN	−2
NICHOLS, WYATT	−1
PURCELL, JACK	−1
ANDERSON, BILL	−1
WRIGHT, ISAAC	−1
FILBERT, MITCH	+1

3. **PHONE NUMBER** What is the probability that a phone number generated using the digits 1, 2, 2, 4, 5, 5, 6, and 2 is the number 654-5222?

4. **LETTERS** Jaclyn bought some decorative letters for a scrapbook project. If she selected a permutation of the letters shown, what is the probability that they would form the word "photography"?

5. **COFFEE BREAK** A group of 6 friends of varying ages meets at a coffee shop and sits in a circle. What is the probability that the youngest member of the group sits in the seat closest to the door?

6. **JEWELRY** Bonita bought her mom a charm bracelet. Each charm is labeled with a one-word message. What is the probability that the 5 charms were hung in the order: dream, believe, love, laugh, inspire?

7. **COLLEGES** Mark wants to visit the 10 colleges he is considering attending. He can only spend the night at 3 of them. What is the probability that he spends a night at Rutgers University, a night at the University of Miami, and a night at Clemson University?

8. **ODD JOBS** Matthew put fliers advertising his lawn service on the doors of 20 families' houses in his neighborhood. If 6 families called him, what is the probability that they were the Thompsons, the Rodriguezes, the Jacksons, the Williamses, the Kryceks, and the Carpenters?

13-3 Skills Practice

Geometric Probability

Point X is chosen at random on $\overline{LP}$. Find the probability of each event.

1. $P(X$ is on $\overline{LN})$

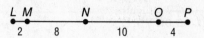

L M   N   O P
2   8   10   4

2. $P(X$ is on $\overline{MO})$

Find the probability that a point chosen at random lies in the shaded region.

3.

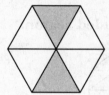

4.

5.

13
5   12   5

6. DESKWORK The diagram shows the top of a student's desk at home. A dart is dropped on the desk. What is the probability the dart lands on the book report?

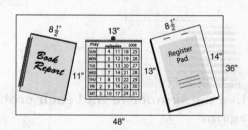

7. FROGS Three frogs are sitting on a 15-foot log. The first two are spaced 5 feet apart and the third frog is 10 feet away from the second one. What is the probability that when a fourth frog hops onto the log it lands between the first two?

8. RADIO CONTEST A radio station is running a contest in which listeners call in when they hear a certain song. The song is 2 minutes 40 seconds long. The radio station promised to play it sometime between noon and 4 P.M. If you tune in to that radio station during that time period, what is the probability the song is playing?

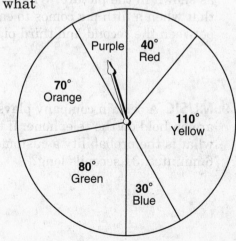

Use the spinner to find each probability. If the spinner lands on a line it is spun again.

9. P(pointer landing on yellow)

10. P(pointer landing on orange)

13-3 Practice

Geometric Probability

Point *L* is chosen at random on $\overline{RS}$. Find the probability of each event.

1. $P(L$ is on $\overline{TV})$

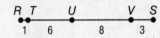
R T   U   V S
1   6   8   3

2. $P(L$ is on $\overline{US})$

Find the probability that a point chosen at random lies in the shaded region.

3.

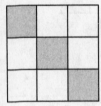

4.

5.

Use the spinner to find each probability. If the spinner lands on a line it is spun again.

6. P(pointer landing on purple)

7. P(pointer landing on red)

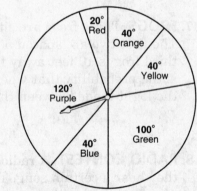

8. **PIGS** Four pigs are lined up at the feeding trough as shown in the picture. What is the probability that when a fifth pig comes to eat it lines up between the second and third pig?

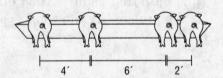

4′   6′   2′

9. **MUSIC** A certain company plays Mozart's *Eine Kleine Nachtmusik* when its customers are on hold on the telephone. If the length of the complete recording is 2 hours long, what is the probability a customer put on hold will hear the Allegro movement which is 6 minutes, 31 seconds long?

13-4 Skills Practice

Simulations

Design and conduct a simulation using a geometric probability model. Then report the results using appropriate numerical and graphical summaries.

1. **INTERNET** Cory has an online store and auction site. Last year he sold 85% of his inventory.

2. **CANDY** Haley works at a candy store. There are 10 types of bulk candy. Find the probability that one type of candy will be chosen more than once in 10 trials.

Design and conduct a simulation using a random number generator. Then report the results using appropriate numerical and graphical summaries.

3. **FOOD** According to a survey by a restaurateur's magazine on favorite types of food, 45% of their readers chose Italian, 25% Mexican, 15% American, 10% French, and 5% Ethnic.

13-4 Practice

Simulations

Design and conduct a simulation using a geometric probability model. Then report the results using appropriate numerical and graphical summaries.

1. **TRACK** Sean successfully handed off his baton 95% of the time in the 4 × 4 relay last season.

2. **BOARD GAME** A game has 50 cards with 10 each numbered 1 to 5, and a player must draw a 2 or a 3 to move out of the "start" position.

Design and conduct a simulation using a random number generator. Then report the results using appropriate numerical and graphical summaries.

3. **REAL ESTATE** A real estate company reviewed last year's purchases to determine trends in sizes of homes purchased. The results are shown in the table.

Homes	Purchase %
2BR	10%
3BR	35%
4BR	30%
5BR	15%
6BR	10%

4. **GRADES** On Jonah's math quizzes last semester he scored an A 80% of the time, a B 15% of the time, and a C 5% of the time.

13-5 Skills Practice

Probabilities of Independent and Dependent Events

Determine whether the events are *independent* or *dependent*. Then find the probability.

1. In a game two dice are tossed and both roll a six.

2. From a standard deck of 52 cards, a king is drawn without replacement. Then a second king is drawn.

3. From a drawer of 8 blue socks and 6 black socks, a blue sock is drawn and not replaced. Then another blue sock is drawn.

Find each probability.

4. A green marble is selected at random from a bag of 4 yellow, 3 green, and 9 blue marbles and not replaced. What is the probability a second marble selected will be green?

5. A die is tossed. If the number rolled is between 2 and 5, inclusive, what is the probability the number rolled is 4?

6. A spinner with the 7 colors of the rainbow is spun. Find the probability that the color spun is blue, given the color is one of the three primary colors.

7. **VENDING** Mina wants to buy a drink from a vending machine. In her pocket are 2 nickels, 3 quarters and 5 dimes. What is the probability she first pulls out a quarter and then another quarter?

8. **ESSAYS** Jeremy's English class is drawing randomly for people to critique their essays. Jeremy draws first and his friend, Brandon, draws second. If there are 20 people in their class, what is the probability they will draw each other's names?

13-5 Practice

Probabilities of Independent and Dependent Events

Determine whether the events are *independent* or *dependent*. Then find the probability.

1. From a bag of 5 red and 6 green marbles, a red marble is drawn and not replaced. Then a green marble is drawn.

2. In a game, you roll an odd number on a die and then spin a spinner with 6 evenly sized spaces numbered 1 to 6 and get an even number.

3. A card is randomly chosen from a standard deck of 52 cards then replaced, and a second card is then chosen. What is the probability that the first card is the ace of hearts and the second card is the ace of diamonds?

Find each probability.

4. A die is tossed. If the number rolled is greater than 2, what is the probability that the number rolled is 3?

5. A black shoe is selected at random from a bin of 6 black shoes and 4 brown shoes and not replaced. What is the probability that a second shoe selected will be black?

6. A spinner with 12 evenly sized sections and numbered 1 to 12 is spun. What is the probability that the number spun is 12 given that the number is even?

7. **GAME** In a game, a spinner with 8 equally sized sections numbered 1 to 8 is spun and a die is tossed. What is the probability of landing on an odd number on the spinner and rolling an even number on the die?

8. **APPROVAL** A survey found that 8 out of 10 parents approved of the new principal's performance. If 4 parents' names are chosen, with replacement, what is the probability they all approve of the principal's performance?

13-6 Skills Practice

Probabilities of Mutually Exclusive Events

Determine whether the events are *mutually exclusive* or *not mutually exclusive*. Then find the probability. Round to the nearest tenth of a percent if necessary.

1. drawing a card from a standard deck and choosing a king or an ace

2. rolling a pair of dice and doubles or a sum of 6 is rolled

3. drawing a two or a heart from a standard deck of 52 cards

4. rolling a pair of dice and a sum of 8 or 12 is rolled

Determine the probability of each event.

5. If the chance of being selected for the student bailiff program is 1 in 200, what is the probability of not being chosen?

6. If you have a 40% chance of making a free throw, what is the probability of missing a free throw?

7. What is the probability of spinning a spinner numbered 1 to 6 and not landing on 5?

8. Jeanie bought 10 raffle tickets. If 250 were sold, what is the probability that one of Jeanie's tickets will not be selected?

13-6 Practice

Probabilities of Mutually Exclusive Events

Determine whether the events are *mutually exclusive* or *not mutually exclusive*. Then find the probability. Round to the nearest hundredth.

1. drawing a card from a standard deck and choosing a 7 or a 10

2. rolling a pair of dice and getting a sum of either 6 or 8

3. selecting a number from a list of integers 1 to 20 and getting a prime or even number

4. drawing a card from a standard deck and getting a queen or a heart

Determine the probability of each event. Round to the nearest hundredth.

5. What is the probability of drawing a card from a standard deck and not choosing an ace?

6. What is the probability of rolling a pair of dice and not rolling the same number?

7. If the chance of being chosen for the principal's task force is 3 in 20, what is the probability of not being chosen?

8. What is the probability of spinning a spinner numbered from 1 to 12 and not landing on 6?

9. **TRAFFIC** If the chance of making a green light at a certain intersection is 35%, what is the probability of arriving when the light is yellow or red?

10. **RAFFLE** Michael bought 50 raffle tickets. If 1000 were sold, what is the probability that one of Michael's tickets will not win?